中式空间

凤凰空间·华南编辑部 编

The Art Of
Chinese Style
Interiors

江苏凤凰科学技术出版社

目录
Contents

传统建筑
Traditional Architecture

茶文化
Tea Culture

禅
Zen

工艺美术
Arts and Crafts

传统建筑

Traditional Architecture

传统建筑

古代中国建筑的历史遗存，覆盖了数千年的中国历史，如汉代的石阙、墓室；南北朝的石窟寺、佛塔；唐代的砖石塔与木构佛殿等等。明清时代的遗构中，更是完整地保存了大量宫殿、园林、寺庙、陵寝与民居建筑群。从中可以看出中国建筑发展演化的历史。同时，中国是一个多民族的国家，藏族的堡寨与喇嘛塔、维吾尔族的土坯建筑、蒙古族的毡帐建筑，以及西南少数民族的竹楼、木造吊脚楼，都是具有地方与民族特色的中国建筑的一部分。

一般来说，中国传统建筑包括官式建筑与民间建筑两大类。官式建筑包括历代皇家宫苑、陵寝，以及敕建的佛寺、道观和坛庙中有斗拱的建筑。民间建筑，除了散落在各地的风格迥异的民居建筑，如士绅商贾的宅院、祠堂、会馆之外，也包括一些地方性庙宇。此外，还有一些不能简单归类的建筑物，如历代建造的佛塔、桥梁、鼓楼、钟楼、市楼等，也都属于中国传统建筑的范畴。

凵 形态特征

中国古代建筑在建筑形态上最显著的特征就是屋顶的飞檐与斗拱，远远挑出的屋檐对屋身的墙体、门窗以及夯土台基都起到了一定的保护作用，重叠的披檐、飞升的翼角及在深远的檐下结构精美的斗，形成了中国建筑的标准母题。

中国古代建筑靠木柱、木梁来承托房屋上部的荷载，因而形成了"墙倒屋不塌"的结构特征，檐下以富有装饰效果的斗拱作为立柱和横梁的连接。斗拱是由弓形的短木块和两层弓木之间相垫的方木块拼合装配而成的特有构件（弓形短木称"拱"，方木形如"斗"），其功用在于承受上部支出的屋檐，将其重量或直接集中到柱上，或间接地先纳至额枋上再转到柱上。在造型上平衡着飞檐的轻巧，使人感到庞大的屋顶在横向的逶迤中趋向稳定。凡是非常重要或带有纪念性质的建筑物，都有斗拱的安置。

我国传统建筑檐部形式之一。屋檐上翘，若飞举之势，常用于亭、台、楼、阁、庙宇、宫殿等建筑上。

斗拱是中国古代建筑独特的构件。从柱顶上的一层层探出成弓形的承重结构叫拱，拱与拱之间垫的方形木块叫斗。两者合称斗拱。

屋顶在中国建筑中虽然有着固定的形式甚至尺寸，但是每个朝代还是变化出了自己的特点：唐代的建筑雄浑大气，出檐平缓，建筑整体偏向横向，而宋朝建筑则变得纤巧秀丽、注重装饰，屋角飞檐有起翘之势，给人以轻灵秀逸的感觉。到了明清，建筑不再侧重结构美，更着重于组合、形体变化及细部装饰等等，所以明清的建筑斗拱层叠、飞檐多装饰、屋顶琉璃瓦金碧辉煌，演化出极具装饰性的形式。

中国古代建筑除了木结构体系的特征外，就是建筑的群体性。单栋房屋体型多为长方形，单纯规整，体量也不大，但是这些简单的房屋却能够组合成住宅、寺庙、宫殿等不同类型和规模的建筑群体。建筑群体的组合几乎都采取院落的形式，通常由四栋房屋围合成院，称为四合院。四合院可以说是中国古代建筑群体组合的基本单元，哪怕是金碧辉煌的紫禁城也是由大大小小的四合院构成，自然也是住宅的主要形式。

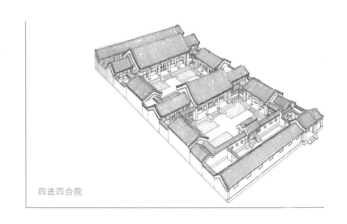

四进四合院

与 建筑装饰特征

中国古代建筑因木结构特征，柱、梁、枋、斗拱等几乎都是显露在外面的，因此装饰要利用各构件的形状，以及材料本身的质感等进行艺术加工，使其构件不仅是构件，也是建筑的装饰语汇。

古代的门，一些看似装饰的附件都与大门的构造有关——成排的门钉是为了将木板与横穿木相连起来，门框横木上角形或花瓣形的门簪其实是固定连楹木与门框的木栓头，而辅首则是为了扣门和拉门。

门钉

门簪

辅首

古代的窗多用纸糊来挡风遮雨，需要较密集的窗格才可，因此，窗格花纹成为了中国窗户的灵魂，各式几何图案、动植物和人物形象被巧妙地融入到边框中。喜好装饰的清代是门窗榥格图案的高峰，与明代简单的井字格、柳条格、枕花格、锦纹格不可同日而语。在清代，许多门窗榥格图案已发展为套叠式，即两种图案相叠加，如十字海棠式、八方套六方式、套龟背锦式等。江南地区还喜欢用夔纹式，并由此演化为乱纹式，更进一步变异为粗纹乱纹结合式样。

古代雕花窗

内檐隔断也是中式装饰的重点，除隔扇门、板壁以外，大量应用罩类以分隔室内空间。仅常见的就有栏杆罩、几腿罩、飞罩、炕罩、圆光罩、八方罩、盘藤罩、花罩等形式，此外尚有博古架、太师壁等。丰富的内檐隔断创造出似隔非隔、空间穿插的内部空间环境。

北京故宫

安徽西递宏村

中国古代建筑装饰的色彩有着固定模式，因为色彩是在伦理的框架内寻求美。唐代以前的建筑主要是材料的本色，到了唐代，建筑一律采用朱红与白色的组合，显得鲜艳、明快。到了宋、金以后，逐渐形成固定的横式——宫殿用白石做台基，墙、柱、门、窗用红色，屋顶用黄、绿、蓝各色的琉璃瓦，在檐下用蓝、绿相配的"冷色"和金碧辉煌的彩画，创造一种绚烂夺目的艺术效果，紫禁城就是最好的例子。而一般民居则是多用白墙、灰瓦和栗、黑、墨绿等色的梁柱和装饰，形成秀丽雅淡的格调。

匕 宫殿建筑

永乐十九年起，明代定都北京，整座城市有外城、内城、皇城、紫禁城（宫城）四重城垣，形成坚实严密的封闭结构。北京城在规划格局上继承了历代传统，以规整对称、突出中轴线的手法，体现皇权尊严。城内的所有宫殿以及重要建筑都贯穿在这条中轴线上，并通过远近开合的手法造成建筑群起伏跌宕、抑扬顿挫之势，又以"工"字形建筑基座、汉白玉栏杆、建筑物的雕梁画栋和黄瓦、红柱、暗红墙面的色彩处理，形成庄严肃穆、宏伟壮丽的气魄。

山 佛教建筑

佛教是中国最重要的宗教，因此佛寺建筑也成了中国古代建筑中重要的建筑类型。佛塔作为纪念性建筑，从印度传到中国后与中国的屋顶形制相结合，形成了两种主要类型：楼阁式和密檐式。楼阁式塔各层面大小与高度都自下而上逐层缩小，整体轮廓呈角锥形，每层辟门窗可以登临。密檐式塔都是第一层特别高，以上骤然变得低矮，层面大小和高度逐渐缩小，越收越紧，各层檐紧密相接，故称为密檐式，整体轮廓呈炮弹形。

中国现存古代建筑并不多，佛殿建筑形制参考了宫殿建筑，而成为了中国古建筑的活标本，梁思成的《中国建筑史》便以大量的佛殿为研究样本，其中以唐代的佛光寺最具代表性。佛光寺大殿有五排立柱，其斗拱硕大，出檐深远，立面宽阔，为佛堂以及信众提供了宽敞的空间。此类雄壮的大殿在都城宫殿中最为常见，所以其被视为唐代建筑的缩影。

山 民居建筑

元大都城的规划产生了胡同与胡同之间的四合院住宅，经过明清两朝，这种住宅近一步得到了发展，于是"北京四合院"成了北京住宅的代名词。北京的四合院院落宽绰疏朗，四面房屋各自独立，彼此之间有游廊联接，起居十分方便。四合院是封闭式的住宅，对外只有一个街门，关起门来自成天地，具有很强的私密性，非常适合独家居住。

中国江南地区住宅在平面布局上同北方的"四合院"大体一致，只是院子较小，称为天井，仅作排水和采光之用（"四水归堂"为当地俗称，意为各屋面内侧坡的雨水都流入天井）。这种住宅第一进院正房常为大厅，院子略开阔，厅多敞口，与天井内外连通。后面几进院的房子多为楼房，天井更深、更小。屋顶铺小青瓦，室内多以石板铺地，以适应江南温湿的气候。

江南水乡住宅往往临水而建，前门通巷，后门临水，每家自有码头，供洗濯、汲水和上下船之用。此种民居布局方式则和四合院相反，不追求整齐划一，不讲究左右对称，因地制宜，布置得体。举凡风景园林、民居房舍以及山村水镇等等，大都采用这种形式，按照山川形势、地理环境和自然的条件等灵活得体布局。

撰文 周航馨

廊坊德发古典家具体验馆

Defa Exhibition Hall
of Classical Furniture, Langfang

名称:

廊坊德发古典家具体验馆

设计公司:

北京凯泰达国际建筑设计咨询有限公司

设计师:

李珂

摄影:

高寒

主要材料:

深灰色地砖、软膜灯箱、丰镇黑烧毛面大理石、深灰色铝板、松木

Name:

Defa Exhibition Hall of Classical Furniture, Langfang

Design Company:

Beijing KaTa International Architectural Consultation CO.,LTD

Designer:

Li Ke

Photographer:

Gao Han

Major Materials:

Dark Grey Floor Tile, Membrane Light Box, Fengzhen Black Marble with Singeing Surface, Dark Grey Aluminium Sheet, Pinewood

这是一个面积近 6 000 平方米的混凝土建筑，一进入大厅，就看到左右对称的 8.4 米宽的白色背景上，有一个放大的圆形木格窗，透着背后的景致。挑空处原来有一条混凝土楼梯径直通向二层，最终这条楼梯被拆除，在挑空的左右两侧新加建了两处楼梯。

Entering into the lobby of this nearly 6,000 m² building, what we can see is a magnified rounded wooden window on the white setting of symmetric 8.4 m piercing the back view. An original stairway on the vertical space which accessed to the second floor was taken down finally, and two new stairways were built up on the both sides of the vertical space.

木格窗
Wooden grille

木格窗是中国传统窗户中的其中一种样式。江浙地区，特别是苏州的工匠，其技巧的精致绮丽也是格外有名的。为克服木材的应力而创造发明的攒插工艺，不仅使几何纹样的门窗展现惊人的细腻风范，也使花节这一原本只是门窗构件的小小物什变得多姿多彩。在中国北方，人们更多的还是用了穿插的工艺，不仅省工省料，而且舒展大方。

圆形的木窗在二层的空间中有序的出现，站在一端望向另一端，层层叠叠，步移景异，时而消失于白墙，时而隐没于虚空，变幻莫测，给人无限的遐想。或许尽头会出现一棵随手在田野摘来的芦苇，抑或是一两朵干枯的野荷，"夫趣得之自然者深，得之学问者浅"。其间味道，自知语言无法尽全其味。暂不论什么是造型，什么是美学，只需懂得生活，便深浅自知，禅意尽显。

The rounded wooden windows which has sequentially appeared in the second floor, are various tier upon tier, and sometimes disappear on the white wall, sometimes conceal in the void. Perhaps a bulrush picked in the field, a sere wild lotus or two will occur in the end, expressing "the pleasure gained from the nature is just deeper than the one gained from learning". Its meaning cannot be fully expressed through the words. No matter what we know about modeling and aesthetics, what we only need to know is life, of which the extents are held by ourselves, and the meaning of Zen will be clear.

整个展厅采取流线分布的方式，每个小空间既自成一景，又融为一体。禅的精髓在于"不说破"，留给他人思考的余地。透过白墙上深色的格栅看背后的光景虚虚实实，影影绰绰，一件件细细打磨的温润木器站在那里，互相遥望，似有低语，似有婉唱。

Each of the small spaces used in the streamline placement of the whole pavilion forms the scenery of its own, but has been combined with others. The quintessence of Zen is "not to reveal", so as to leave the room for others to think about. You can make a glance for the back view through the deep-colored grille on the white wall to enjoy the delicacy of carpentry. The mildly-polished carpentries there seem to look at each other, talk with each other, and even sing gracefully.

建筑选材力求质朴平和，用物质上的"少"去寻求精神上的"多"。混凝土的顶和暴露的管道喷成了白色，使空间更加浑然一体。大厅中的《富春山居图》背景，在深色木格栅的掩映下静静的散发着柔和的光晕，黑灰色的地面，隐隐反射着光、影、景。《无用师卷》的布局、笔墨、意象与我们所追寻的禅意相得益彰。

《富春山居图》
Dwelling
in the Fuchun Mountains

《富春山居图》是"元四家之首"黄公望晚年的水墨山水巨作，中国文人山水画的典范。被誉为"中国十大传世名画"之一。它用长卷的形式，描绘了富春江两岸初秋的秀丽景色，把浩渺连绵的江南山水表现得淋漓尽致，达到了"山川浑厚，草木华滋"的境界，被誉为"画中之兰亭序"。该画后历经磨难，分为两个部分。"剩山图"今藏于浙江省博物馆，而"无用师卷"则藏于台北故宫博物院。

竹藤骨架的圆形吊灯错落悬垂其间，光线温和柔软，楼梯下方两汪清潭静静的倒影着山水，清潭里时而游过几尾红鲤，搅动水波，让光影四散开去。"空山鸟语兮，人与白云栖，潺潺清泉濯我心，潭深鱼儿戏"，含蓄、单纯、自然，才是设计所要表达的美。坐在大厅的山水高墙下，品一壶茶，手中木器的触感、品相仿佛带上了时光的味道，大有脱迹尘纷之感。

The building materials look plain and mild in the use of them, while we have used the material "less" to seek the spiritual "more". The top of concrete coiling and exposed pipeline have been sprayed to the white color to make the space become one harmonious whole. The *Dwelling in the Fuchun Mountains* in the lobby is exuding the soft halo quietly at the back of deep-colored wooden grille. The layout, skills and image of Painting for *Wuyongshi* (the second part of *Dwelling in the Fuchun Mountains*) just suit the Zen's real meaning we are seeking. The black grey floor is faintly reflecting the light, shadow and scene.

The round pendant lamps with bamboo and rattan framework are overhanging in picturesque disorder. With the mild and soft light, landscape is reflected quietly on the two pools under the stairway. Several red carps swimming in the pools are stirring the water to scatter the light and shadow away. "The birds are singing in the valley, as we are having a rest with the clouds. The gurgling spring is cleaning my soul, while the fish are playing in the deep pond". Connotation, purity, nature, is just the beauty which the design is expressing. While sitting below the high wall with landscape, and sipping a cup of tea, the tactility and condition of carpentry on the hands seem to have brought the flavor of time, and the sense of detaching from the world.

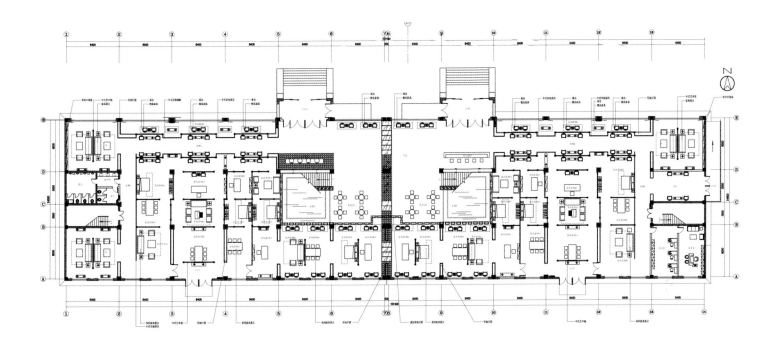

一层平面图
First Floor Plan

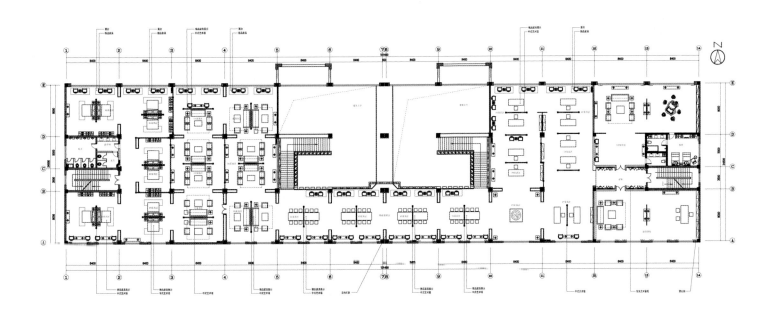

二层平面图
Second Floor Plan

茗古园·金丝楠木汇馆

Silkwood Clubhouse,
Ming Gu Garden

名称:

茗古园·金丝楠木汇馆

设计公司:

品川设计

设计师:

陈杰

摄影:

周跃东

撰文:

全剑清

Name:

Silkwood Clubhouse, Ming Gu Garden

Design Company:

PURE'CHARM SPACE DESIGN ORGANIZATION

Designer:

Chen Jie

Photographer:

Zhou Yuedong

Artucle:

Quan Jianqing

夜幕时分，店面上方的浮云图案泛着光泽，散发着一种若即若离的气息。汇馆的外立面用透明的玻璃材质，让内部空间的景致成为一张鲜明的视觉名片，哪怕是瞬间的吸引，这种感知都被记忆在人们的生活影集里。

In the nighttime, the cloud patterns embedded on the brand board are sparkling, creating a charming atmosphere. The exteriors of the clubhouse are designed with transparent glass materials, which allow the viewers to see through the interiors from the outside. In this way, a visual name card is instantly established and stays in the viewer's mind.

古朴的空间不仅是视觉的体验，更是一种发自内心的愉悦。茗古园·金丝楠木汇馆便是这样一个场所，空间中的家具以及陈设都是灵感的汇合，它们以典雅的色彩和线条诠释着中国传统文化的精髓，并以清茶般的苦后回甘来表达自身的韵味。进入其中，我们仿佛走入另一重境界，身心不自觉地便摇曳在艺术与文明的氤氲情境之中。

A rustic space is not only a visual experience, but also the joy stemming from within. The way the furniture are laying out in Silkwood Clubhouse represent the essence of

Chinese traditional culture with its elegant color and vintage lines. Entering here, we are like walking into another level of life, and the whole body and mind would subconsciously swing among the environment of arts and culture.

❀ 唐代仕女
Tang Dynasty Ladies

"仕女"一词最早见于唐朝。古代称做官的人为"仕"，"仕女"泛指上层社会的妇女，特别是封建王朝时期的贵族妇女，代表的是中国古代那些美丽聪慧的女子，也是历代画家热衷描绘的对象，中国画中还有专门的"仕女画"类别。唐代作为封建社会最为辉煌的时代，也是仕女画的繁荣兴盛阶段，多展现唐代贵族妇女生活情调。

The gate of the clubhouse is designed to be like a round hole in the wall, the visual connection between the inside and the outside makes the house extremely inviting, unconsciously expressing the traditional culture. The wall of the entryway is decorated with various Chinese architectural accessories. These unique objects with different texture and patterns greatly enrich the wall covering and bring in some delicate and déjà-vu experience.

汇馆的门以园洞的形式存在，内外的视觉衔接让空间拥有了十足的亲和力，并潜移默化地将传统意味铺陈开来。进门后的玄关墙以各式中式建筑构件作为装饰，这些构件模糊了时间的概念，却颇有一番自在的个性，有着各自不同的纹理质感。层层叠叠的组合丰富了墙面的层次，带来了微妙又似曾相识的体验。

在汇馆里或走或停，人们的思绪不会出现断层，因为材质与情调在这里融为一体。设计师用淡定从容的细节主张，描绘传统生活的缩影，看似从感官上的喧嚣中回归朴实无华，实则拉开了一幕精彩的篇章。

When people are walking around here, they would pause in thinking and feel like blending in with the materials and atmosphere among the space. The calm and easy details created here by the designer are all reflections of a traditional lifestyle. The simple and natural decorating here is in fact the opening of an exquisite feast.

走道的右侧是汇馆的主题空间，上方的装饰延续着老建筑楼梯构件的装饰，向下的结构倾向使得人们的视觉焦点自然而然地落到展示区域中的金丝楠木家具上。

这个空间展示以金丝楠木新料为主制作的家具，这些家具的设计历经几千年的文化洗礼，至今依然风姿绰约。它的存在诠释着一种风情，代表了一种文明。

The right side of the path is the theme space of the house. The decorating on the ceiling exquisitely applied some ancient ladder elements with their downward structure, and it naturally focuses people's view onto the silk wood furniture on the ground.

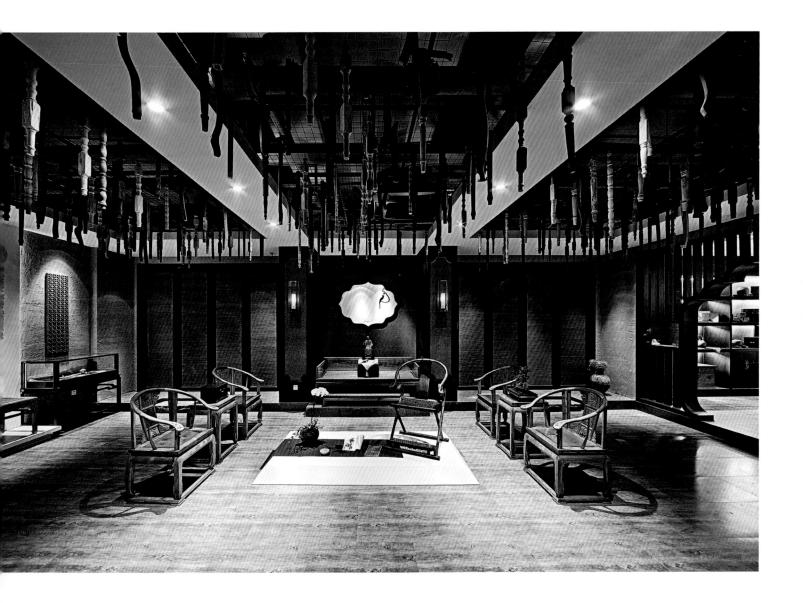

This space aims at displaying its furniture mainly manufactured from silk wood. After thousands of years of culture alteration, this unique furniture beautifully stand still among China's traditional culture.

这个空间的背后设置了一个以金丝楠老料为主的书房家具展示空间，里面的每一件家具、器皿都像老朋友一样，用不着过多的语言渲染，便能用其姿态向我们诉说其存在的缘由。置身其中，仿佛能听到他们在讲述一个个精彩的故事，让我们彼此更加接近。与之毗邻的房间，除了金丝楠木家具外，还陈列了其他材质的收藏品，这些珍奇的收藏，都与主人有着这样或那样的缘分。

There is a show room of the study and its furniture set behind the clubhouse. The furniture here is mostly made from silk wood. Each and every piece of furniture are like old friend of the owner. Much is being said here with few words, since each existing piece is like telling a wonderful story that connects us to the room. Besides the silk wood furniture, there is many other novel collections of the owner displayed here, attracting people to come round and linger in the space.

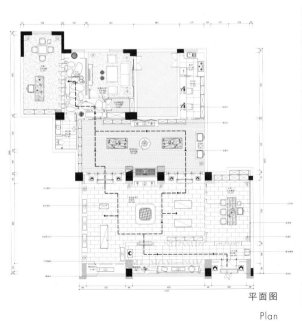

平面图
Plan

梓山湖领御营销中心

Marketing Center
of Zishanhu Lingyu Garden

名称：

梓山湖领御营销中心

设计公司：

柒零捌零内建筑设计事务所

设计师：

胡勤斌

摄影：

江河摄影

主要材料：

泰顺青石 榆木 玻璃 不锈钢 油漆

Name:

Marketing Center of Zishanhu Lingyu Garden

Design Company:

7080 Interior Architecture Design Studios

Designer:

Hu Qinbin

Photographer:

Jiang He Photo

Major Materials:

Taishun Bluestone, Elm, Glass, Stainless Steel, Paint

本案萃取淳朴、秀美的江南水乡精髓，以其从容淡泊的水墨意象在城中央独树一帜。没有赘余的装饰，亲近平和的中式风格建筑会让您忆起悠悠往昔岁月。

The Project represents the simple, clear essence of Yangtze River Delta and humble image of Ink Painting, to be unique in the center of the city. Without any superfluous adornment, the warming Chinese-style building will remind you of the past years.

紧密围绕"山水、人文、生态"与"花鸟、书香、茶品"等元素构筑一个富贵和谐的生活氛围，使空间更显生气与灵性。

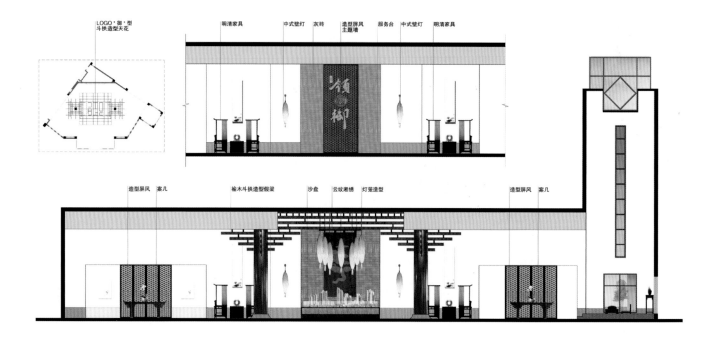

The rich and harmonious living environment which is built up tightly around some Chinese elements such as "Scenery, Humanity, Ecology" and "Flowers & Birds, Books, Tea", makes the space much more lively and spiritual.

朴实的钢构线条月洞门及中式传统门框配以现代自动化玻璃门、木制屏风、祥云宫灯、明清案几、青石台与书架，层层递进、步步取景，创造一种东方水墨印象。

The unadorned steel line moom-shaped hole, Chinese traditional door frame with modern automatic glass door, wooden screen, palace lanterns and Xiangyun (auspicious clouds), tables from the Ming and Qing Dynasties, bluestone desks and bookshelves, all of the views are step by step, creating the oriental ink-painting impression.

月门
MOON GATE

即月洞门，用以形容开在院墙上，圆形如月的门洞。月洞门左右延伸出去的廊房，向内环抱着酒泉，飞檐彩绘，古雅宁静，是典型的明清风格。月洞门是中国古代园林必不可少的象征元素，有园林，必有月洞门。它不但成为进入园林的重要通道，也是了解中国园林文化的特殊符码。

材料：室内延续建筑中白墙灰瓦的灰白主色调，以项目 LOGO 为装饰主题，一进入大厅就设置了一个中式木格屏风，依旧是以中式木格和月洞门为设计元素，与大门的造型相呼应。以轴线一字穿插的主题背景墙，采用木质花格为底，镶嵌钢制项目 LOGO，古与今的材质结合，更能体现新中式装饰风格在细节上的创新。

Materials: The interior continues the grayish white dominant tone of the white and gray tiled building. With the Project logo as the adornment theme, the Chinese-style wooden lattice screen and arch echo with the modeling of the moon gate. The theme wall going straight through the axis, has been replaced by the wooden lattice theme wall, while the ancient materials of the Project logo in right proportion are combined with the modern ones to better reflect the innovation in details of new Chinese style decorating.

此案的设计重点是营销中心的互动区（此区域成本占项目的大部分），项目总成本为 150 万，综合考虑成本问题，木作方面用的是价钱合理的榆木。天花，以提炼后的中式斗拱为元素，由荔枝面灰麻基底榆木柱做支撑，简化的现代斗拱，围绕中心的一个"御"字展开做造型，御字象征中国贵族之权威，加上排列有序的祥云图案吊灯，气势非凡。

The key point of the Project is the interactive area of the marketing center (of which the cost takes up most part of total cost of the Project). As the total cost is 1,500,000, as a material with reasonable prece, elm has been widely used in the carpentry. First is the ceiling which represents the essence of the traditional Chinese dougong element and is supported by elm pillars upon the gray rough base, while the modern modeling of dougong embraced the ceiling shaped as the character of " 御 (Yu)", which symbolizes the power of China's nobility. The orderly pendant lights in the auspicious cloud pattern are of great magnificence.

整个销售大厅地面用的是产自温州泰顺的青石砖，以工字铺法展开，为我们想表达的新中式铺垫了基础。墙面以白色为基调，墙裙及门框线则特意挑选色差大、手工精细的灰砖，通过这些元素隐性植入整个案例，很好的营造一种古色古香而又不失清新雅致的新中式氛围。

The bluestone bricks from Taishun, Wenzhou have been used on the ground of the marketing hall in I shape to lay foundation for the new Chinese style we want to express. The large-chromatic and fine-handwork gray bricks are selected on some parts of the extensive white wall, the dado made of gray stone and door frame line have been used. The whole case has been implicitly absorbed through these elements to create the antique and graceful new Chinese-style ambiance.

陶艺：青花陶缸、白玉瓶、白梅瓶、红荷花彩陶、或工笔写意、或浓墨重彩，将水乡文化体现得淋漓尽致。

Potteries: There are blue-and-white earthen jar, white jade vase, white prunus vase, painted potteries with red lotus, etc. All these artworks vividly reflect the culture of the local town. Their nice modeling not only meet the aesthetic taste, but also embodies a new conception fused with the traditional culture.

黑茶：16世纪末期，湖南黑茶兴起，湖南黑茶原产于安化，《茶马驿馆咏》一文中有所介绍，旨在弘扬不畏艰险的千年茶马精神。此案融入了当地人比较熟悉的黑茶文化，在高耸的壁橱内放置黑茶，并设计了专门的茶艺区，以便与客户有更进一步的交流。

Dark Tea: Originating in Anhua, Hunan, Dark Tea has risen from the 16th century. It is introduced in the article *Ode to the Tea-horse Courier Hostel* to promote the fearless ancient tea route spirit that lasts for over a thousand years. The Project catered to the culture of Dark Tea that the local people know well. The Dark Tea has been placed into the towering closet while the specialized tea art area has been designed functionally to make further communication with the customers.

❋ 折枝
FLOWERS IN BRANCH

"折枝"有二意：1. 择取草茎树枝，比喻轻而易举。后引申为愿为长辈效劳之意。2. 花卉画法之一。

"折枝"是中国花鸟画的表现形式之一，画花卉不画全株，只选择其中一枝或若干小枝入画，所以得名。折枝花鸟的形式出现在在中唐至晚唐之际，宋元时，折枝画法已较为普遍，明清更为盛行。扇面之类的小品花卉画，往往以简单折枝经营构图，弥觉隽雅。

功能划分： 借助隔断巧妙地将营销中心划分为迎宾区、互动区、前台区、VIP区、休憩区等，借助区域功能属性强化售楼服务界面，使客户从进门开始便能享受到最温馨的服务直至离开时仍有"流连忘返"的感觉，一改以往给人的那种功利性强、目的单一的营销攻防氛围。

Functional Division: Through the partition, the marketing center has been tactfully divided into the welcome area, interaction area, front desk reception area, VIP area, recreation area, etc. The sales service interface has been strengthened through functional distribution of area,making the customers enjoy the coziest services inside the door, and still feeling nostalgic when leaving. These can change people's image of the utilitarian, direct and monotone marketing atmosphere in the past.

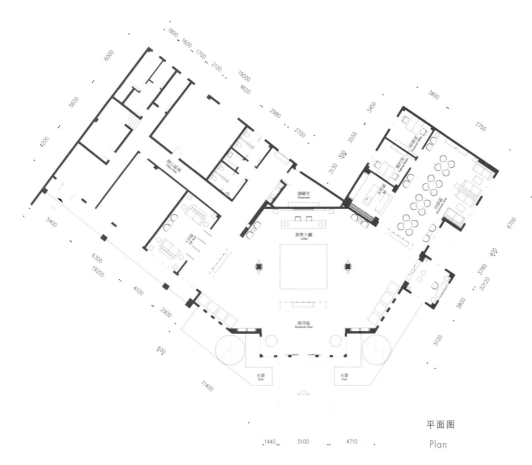

平面图
Plan

东方大院家具展厅

Oriental Yard
Furniture Exhibition Hall

名称:

东方大院家具展厅

设计公司:

道和设计机构

设计师:

高雄

摄影:

周跃东

主要材料:

黑金龙大理石、直白纹大理石、雅士白大理石、白色烤漆玻璃、
黑色不锈钢、乳化玻璃

Name:

Oriental Yard Furniture Exhibition Hall

Design Company:

DAOHE DESIGN

Designer:

Gao Xiong

Photographer:

Zhou Yuedong

Major Materials:

Black Gold Dragon marble, White marble, Aston white marble,
White paint glass, Black stainless steel, Emulsification glass

不同质感的色彩被鲜明地铺洒在整个空间内，中式元素经过提炼后分撒在各个角落，既不是忸怩的，又不是直坦坦的，耐人寻味。空间中没有刻意的渲染，也没有繁缛的修饰，一切看上去都简简单单却又不失凝练。这种对空间的理解与再创作，是来自于设计师内心的淡然和喜悦，并在空间中一层层地传递开来。

All space is scattered with bright colours of different textures and Chinese elements are dispersed at every corner. There aren't deliberately renderings or multifarious embellishments in the space. All of them look very simple but compact. The understanding and redesign of the space come from the calmness and delight of designer.

充满艺术感的彩色琉璃吊灯和不锈钢材质的运用，打破了传统中式的固有模式，创造了设计师自己独特的设计语言，会客厅工笔花鸟画和局部的蓝色运用，更衬托出家具的古香古色。

The artsy decorative lighting and the stainless steel materials endow the Chinese space with modern gene. It breaks the traditional stereotype and creates designer's own unique design language.

Chinese Fine-Brushwork Painting for Flower and Bird and the segmentation of blue wall set off the antique beauty of furniture.

❀ 工笔花鸟
Chinese Fine-Brushwork Painting for Flower and Bird

工笔花鸟画是在专用的熟宣纸或矾绢上进行严谨精致创作的一种特殊画种与技法，是相对于写意花鸟画的一个重要的流派。它形成于唐，成熟于五代，而兴盛于两宋。其中到了五代，以黄荃和徐熙为代表的两大画派，世称"皇家富贵"、"徐熙野逸"，所创立的"勾填法"、"勾勒法"、"没骨法"等画法使花鸟画的表现达到了一个新的艺术水平，标志着工笔花鸟画的技法已经趋于成熟。

工笔花鸟画，表现方法工整细致，先勾后染，设色艳丽，富有装饰性。在对花鸟描绘的过程中，通过白描造型、勾勒填彩，再采用分染、罩染、统染、点染、接染、撞水、碰色等技法描绘对象，产生栩栩如生、精致动人的视觉效果。

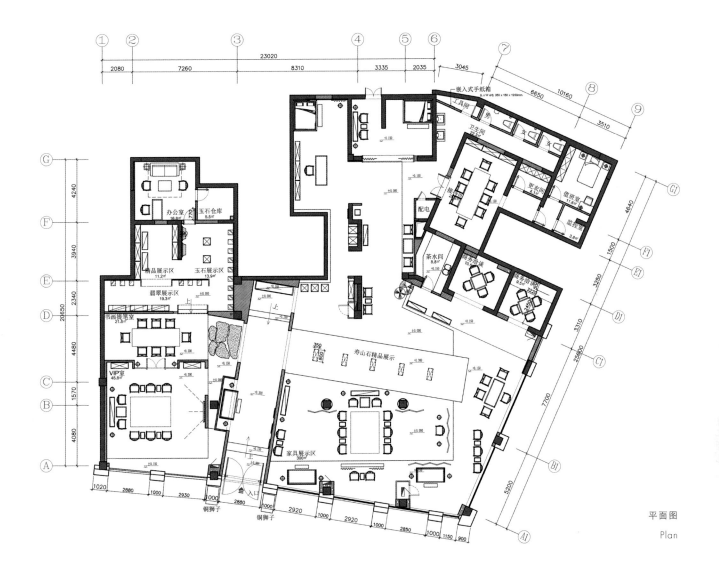

平面图
Plan

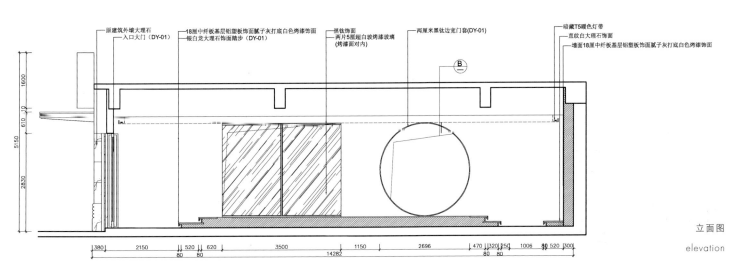

立面图
elevation

东地素舍微酒店

Dongdi Sushe
Mini Hotel

名称:

东地素舍微酒店

设计公司:

柒零捌零内建筑设计事务所

设计师:

胡勤斌、钟东伟

摄影:

江河摄影

主要材料:

青石、榆木、玻璃、油漆、木雕等

Name:

Dongdi Sushe Mini Hotel

Design Company:

7080 Interior Architecture Design Studios

Designers:

Hu Qinbin, Zhong Dongwei

Photographer:

Jiang He Photo

Major Materials:

Bluestone, Elm, Glass, Paint, Wood Curving, etc.

豪华套间的设计风格独特，将素舍的美演绎到了极致。设计师在设计上尊重民居的原有结构，保持院落的传统精神，在设施方面则有限度地融入新世纪的国际标准，体现出现代生活风尚。自然材质的选用旨在营造一种既精致又闲适的氛围，设计师极具创意地把海南三亚细沙布置在客房内，让您足不出户也能享受到"椰林树影、水清沙幼"的热带风情。此外，在主题客房中还添设了私人影院，同时也是KTV包间，给您带来多方位的感观享受。

The design of the deluxe suites show the unique style and ultimate beauty of the plain house. We respect the dwellings' original structures, and keep the traditional spirit of the courtyard. Meanwhile we have also integrated the international standards to embody the modern life style. The use of natural material aims to create a delicate and leisurely ambiance: the silver sand from Sanya, Hainan has been creatively laid out in the guestroom to make you enjoy the tropical temptation of "coconut grove, clear water with soft sand". Furthermore, the private cinema has been added into the theme guestroom, which can also act as KTV room. It can bring you the all-round sensory enjoyment.

项目是由九个结构独立、大小不一、交错相连的院落组成，这些建筑群体现出了中国民居的本土特色。强化中心轴线，以精心设计的中轴串联不同功能的建筑群体，即便是酒店内部也是以现代中式风格来延续历史及文化的脉络。中轴线上的主入口以"拱门"形式来设计，引人入胜，令人仿佛进入桃源般的心旷神怡，一下从喧嚣的外部都市过渡至幽静的内部酒店空间。庭院中的趣味雕塑和流水辅之以传统青石铺就的过道，防腐木制的台阶及LED灯勾勒的玻璃拦河，将不同的院落巧妙地连接起来。夜晚时分，华灯闪烁，令人体验到时光交错，昔日重来的感觉。

The Project is formed by nine compounds with the independent and different exteriors. These buildings reflect the home feature of Chinese dwellings. We have strengthened the axle wire through which the buildings with different functions are connected in series. The modern Chinese style will also be used to continue the vein of local history and culture even if in the hotel interior. The main entrance on the axle wire in an attractive form of "arched door" can make you feel relaxed as if entering into Shangri-la and moving from the blatant city to the peaceful space in the hotel. The interesting sculpture, running water in the yard, supplemented by the aisle paved with bluestone, steps of antiseptic wood, balustrade of the LED light, link the different compounds skillfully. At night, the sparkling light will make you get the feeling of "time staggering, yesterday once more".

雕塑是由设计师与北京雕塑艺术家道明先生经过多次讨论，最终呈现出来的一个现代与趣味相结合的艺术品。与周边环境融为一体，实现了现代时尚元素与古朴民居的完美结合。

The interesting sculpture that combined modernity and quirkiness has been designed after the designer's repeated discussions with Mr. Daoming, a sculpture artist in Beijing.

屋顶材质采用的是透明玻璃与原木，阳光透过玻璃，在白色的砖墙上洒下点点阳光，斑驳中略带温暖。复古家具、开放式吧台，以及供自由阅读的书架，屋内的摆设有一种信手拈来的优雅与舒适，似乎在召唤着渴望宁静的人们。

Clear glass and raw wood used for the roof, to spill a plume of sunlight on the white brick wall. The vintage furniture, opening bar, bookshelf for reading freely...The furnishings in the house express some sense of elegance and comfort, and seem to be attracting the people's eager for serenity.

"东地素舍·茶艺室"传承中国古典意象，在红木家具与水墨书画装点的空间中，三五知己，五六道茶，谈天说地，闲坐其中，只为体会佳茗的幽香。此外，茶艺室也是一个多功能厅，可以满足14人以下的聚会及餐宴。优雅的自然光线里，或品茶论道，或把酒言欢，物我两忘于其中。

"Dongdi Sushe-Tea room" inherits the Chinese classical image. In such a space decorated by mahogany furniture and ink painting and calligraphy, you can sit at leisure, talking with your friends, and tasting the delicate fragrance of tea. Moreover, the tea room is a multifunctional room which can be used to hold meeting and gathering under 14 people. You can take a sip of the tea or take up the wine cup and chat merrily, enjoying yourself amid the elegant natural light.

❀ 岭南民居
The Lingnan Dwellings

由于岭南气候炎热，风雨常至，这里的民居形成了小天井大进深，布局紧凑的平面形式，既便于防暑防雨，又可以通风散热。同时岭南民居由于民系众多，历史悠久，对外交流频繁，形成了千姿百态的民居形式，有城镇型民居，如广州西关大屋，有广府的三间两廊式，有潮汕式民居，有客家民居，有侨乡碉楼民居等。

私房菜馆设有中式包厢和日本榻榻米式包厢，典雅、私密的用餐环境，与独家秘制的私房菜相辅相成。能俯瞰整个下坝坊的楼顶餐厅是附近海拔最高点，凭栏远望，四周都是错落有致、透着古灰色调的旧民居。和钢筋水泥的都市带给你的压抑感不同，这里是夹杂着人间烟火味儿的闲适自在。

The private cuisine restaurants have Chinese-style and Japanese tatami box rooms. The elegant and private dining environment corresponds with the distinctive delicacies. Looking over the fence of the roof restaurant which is the highest place in the whole Xiabafang, there are old short dwellings with the ancient gray tone. Different with the oppression of urban buildings, this is full of leisure and comfort of common living.

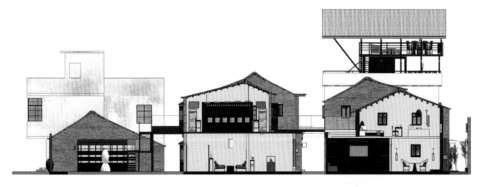

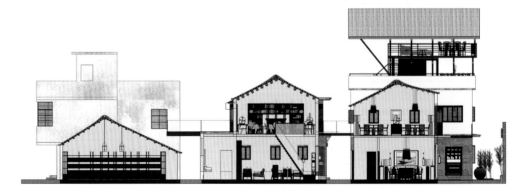

一层平面图

First Floor Plan

二层平面图

Second Floor Plan

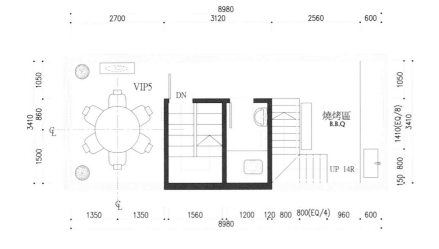

三层平面图
Third Floor Plan

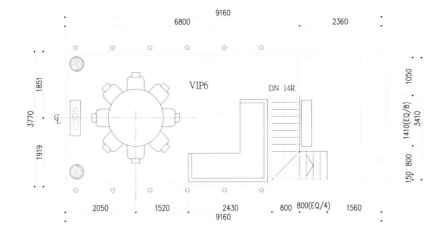

四层平面图
Fourth Floor Plan

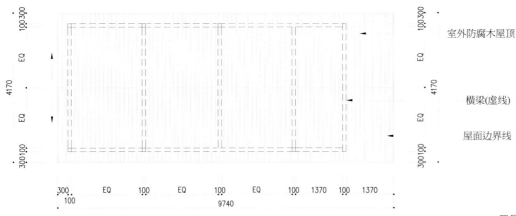

室外防腐木屋顶

横梁(虚线)

屋面边界线

顶层平面图
Roof Floor Plan

改建项目对于城市发展来说，关系着现代生活的渗透以及无处不在的历史文化痕迹，前者意味着城市必然不断有新的物质和功能要素的介入，后者包含着更广泛的内容，如城市的景观和风貌、尺度和肌理等。而改造性再利用的关键是对重要结构的改变降低到最低限度，保存和再现建筑原有的重要性，让本身散发着神秘色彩的老房子更增添一份魅力。既有古典意境的传承，又有现代风格的凸显，身临其中，感观无时不愉悦。一个精致的休闲之所、一个清雅的艺术之境、一颗传统与现代共同孕育的明珠。臻于细节、卓于内涵的东地素舍，将为您揭开东莞生活的另一面——融汇着艺术创意与个性享受的休闲生活。

The reconstruction project is related to the infiltration of modern life and ubiquitous historic cultural marks for the city development. The former one means that the new materials and functional elements have to get involved in the city, while the latter contains much more extensive details, such as the city's scenery and features, scales and textures, etc. The key to adaptive reuse is to lower the change of important structure to the minimum, so as to keep and maintain the buildings' original mystery and glamor. There is not only the heritage of classic meaning, but also the highlight of modern style. Being around such honses, the sensory is full of pleasure. An exquisite place for leisure, an elegant place of art, a pearl conceived by the tradition and modernity, a detail oriented, connotative Dongdi Sushe, will uncover the other side of life in Dongguan for you-the casual living with collision of art and enjoyment.

海联汇

Hai Lian Hui

名称：

海联汇

设计公司：

福州宽北装饰设计有限公司

设计师：

郑杨辉

摄影：

周跃东

撰文：

朱含薇

Name:

Hai Lian Hui

Design Company:

Fuzhou COMEBER Design

Designers:

Zheng Yanghui

Photographer:

Zhou Yuedong

Article:

Zhu Hanwei

位于塔头街 1 号的海联汇是一间重新翻修，呈现混搭新东方风格的餐厅。作为海联酒店的配套项目，业主希望餐厅的重装既要沿用原酒店大堂空间的暖色调，又不乏东方文化的意境。基于此，设计师将现代感和东方意境的两种元素混搭，在材质的穿插、空间体块的组合、造型的意象模拟和软装陈设的百变拼盘中传达了大气深邃的东方意境，非常符合当下的审美特点。

Located in No. 1 Tatou Street, Hai Lian Hui is a renovated restaurant that presents the new oriental mix-and-match style. As a related project of Hailian Hotel, the renovation of the restaurant will not only follow the warm-colored tone in the original hotel lobby, but also be filled with the prospect of oriental culture. On account of this, the designer matched modernity with oriental elements to convey the deep, high-class oriental prospect and current aesthetic features in cross of textures, combination of space blocks, formative image simulation and display changeable assorted plates.

设计师从"以国际性的视野，做区域性的文化"的理念出发，将代表老福州文化的装饰元素巧妙运用到设计当中。餐厅入口处的背景墙以描绘着昔日大洋百货、中亭街、仓山老城区的素描为装饰。包间内部则被以三坊七巷的旧建筑符号为题材定制的手绘作品占据半壁江山。餐厅外部入口的青竹、泉水、陶缸、荷叶让这个空间看起来十分的清幽雅致。入口内侧还开辟了品茗区，以明式家具的硬朗造型传达古朴、闲适的氛围。

Based on the designer's idea of "making the regional culture with the global vision", the decoration standing for the old Fuzhou culture is used skillfully in the design: The Grand Ocean Department Store, Zhongting Street, Cangshan's old town in the past have been painted on the background wall in the restaurant's entrance. In the interior parlors, the hand-painted works with the theme of old buildings take up half of the paintings. The restaurant's exterior entrance sends forth a light leisure and elegance with the bamboo, spring water, lotus leaves and the earthen jars. The tea-tasting area has been opened up in the internal entrance, with the tough shapes of Ming-style furniture to show the rustic and relaxed ambiance.

❀ 水墨画
INK WASH PAINTING

水墨画：经过调配水和墨的浓度，由水和墨所画出的画，是绘画的一种形式。水墨画，更多的时候是中国绘画的代表，也就是狭义的"国画"。基本的水墨画，仅有水与墨，黑色与白色，但进阶的水墨画也有工笔花鸟画，色彩缤纷，后者有时也称为彩墨画。据画史记载水墨画始于唐，成熟于宋，兴盛于元，明清以后有进一步发展。

唐以前的寺、馆、庙、院、殿、窟流传下来的画作，属于重彩勾线填色的作品，水墨画的踪影未现，其形成与发展受道家哲学影响。道家鼻祖老子认为"五色令人目盲"，主张朴素美。唐代王洽首创泼墨画，王洽泼墨画的出现，开创了泼墨画的先河，其技法与用水墨作为表现形式的水墨画有异曲同工之妙，极大地影响了后代的水墨画发展。明清以来，形成了文人画流派，讲究墨色层次变化，并将书法用笔渗于画中，强调水墨索静框架，成为中国画的基本特征。但这些文人画也有不足的一面：追古摹临，缺少创新，使水墨画走向陈腐的道路。在民国时期，文人画受到了众多有识之士的批判。近年来，为了重塑中国绘画在国际艺坛中的地位，当代水墨越发受到重视。

新东方风格的巧妙之处在于将传统意境与当代艺术、传统元素与当代手法巧妙融合，在本案中，设计师恰如其分地表达了这别有韵味的复合式美学。空间的结构通过大体块的拼接构成，用块面衔接的方式穿透延伸。无论是以传统"弓"字形护栏做隔断的公共就餐区，还是用现代玻璃、方钢和木质踏板围合出的透明包厢，抑或是以原木板打造的隔断屏风、书架和陈列柜，不同体块之间的组合刻意而又自然，构成了极具表现力的功能区域。空间布局讲究每一个细节的搭配，每个单品对于风格的表现都有其自身的意义。

The ingeniousness of the New Orientalism is just the blend of traditional prospect and modern art, traditional elements and contemporary techniques. The designer has appropriately expressed this combinated aesthetics with a great deal of charm. The spatial structure is jointed with the large blocks, penetrated and extended by lapping with the block surface. The combination of different blocks is deliberate and natural to form the functional area full of expression, either the public dining area separated by the curving barrier, transparent rooms enclosed with modern glass, square steel and wooden pedal, or the partition screens, bookshelves, showcases made of the log blocks. The spatial layout pays attention to the matching of every detail, while each of the items has its own meaning for the integrity of style.

设计师根据业主的要求，将"水"和"海"的概念作为餐厅设计的主题，以水波的圆弧纹理为灵感，将形态、大小、组合方式不一的"圆"呈现于空间各处。

The designer has put the idea of "water" and "sea" as the theme of restaurant's design according to the owners' requirement.

In all of the elements symbolizing sea and water, the designer presented the "circle" in different forms, sizes and combinations throughout the space, with the inspiration of water wave.

餐厅外立面墙上错落镶嵌的各式圆形陶盘装饰，质朴之余，也在传达关于餐饮的信息。大面积的天花和背景墙以圆弧为肌理，空调出风口则设计成水纹状的圆形，犹如水波荡漾。入口及大包厢的玻璃表面漆上像海水的螺旋纹，大圆套小圆的效果被不断重复。公共就餐区的隔断围栏内，白色水管织成有序的纵向线条，传达"雨"的概念。

The circle terrines on the exterior façade and in the restaurant are plain, conveying the information about catering. The arc shapes have been brushed on a large area of ceiling and background wall, and the air conditioner's outlet is designed into a circle like the water wave. The seawater spiral shape has been painted on the glass surface of the entrance and large balconies, the effect of great circle covering the small one has also been repeated constantly. The partition fence in the public dining area was formed by the white leveling pipes in well-ordered vertical lines to convey the idea of rain.

以水母、海藻等海底生物为原型设计的落地灯散落在餐厅各个空间，其婀娜的体态和流水般的纹理，风姿绰约撩人心弦。各式造型独特的藤制工艺品穿插其中，在无形中又为空间增添了几分闲情野趣。

Moreover, the floor lamps created from the prototype of the marine benthos like jellyfish and seaweed have scattered throughout the space. With the winding postures, streamy textures, and charming integral shapes, rathan artcrafts have invisibly added some simple and elegant taste to the space.

有别于一般新东方风格的华丽和复古，海联汇更讲究通过陈设与配置来营造商务空间中的人文气息，着重于空间品位的提升，在精简传统中式元素的同时，又不失东方意境。我们的视野里并未出现中式装饰中常用的木刻雕花、青花纹理、大红灯笼，但就是那一抹清泉、一张藤椅、一个蒲团让我们感受到其内敛的中式禅意。

Different from the gorgeous feeling and vintage feature of the new oriental style design in general meaning, Hailianhui pays more attention to display and allocation to create cultural atmosphere in the business space and focuses on uplift the taste. The oriental prospect hasn't been lost while the traditional Chinese-style elements are simplified. The wood carving, blue-and-white texture, red large lanterns frequently used in the Chinese-style decoration are not seen in our vision, but there are just the clear spring, a rattan chair, a piece of pouf which make us feel their reaerved Chinese-style Zen.

由各种宽度不一的原木组合而成的屏风上刻着唐代古诗《春江花月夜》的节选，原木书架和陈列柜里分别放置了陶瓷工艺品和传达药膳养生概念的中药药材及各种干菌菇。以酒文化为主题的包间，墙壁层板上摆满了葡萄酒。几张经过抽象拉伸处理的藤制餐椅、印象派风格的水墨画作等现代元素传递出更加多元的审美主张。当意象化的东方元素与现代时尚感邂逅，当那些象征古韵的气息悄然融入现代工艺，包容、内敛而低调神秘的混搭新东方风格便呈现了出来。

Excerpts from a Tang's poem *A Wonderful Night in Spring* are carved on the screen combined with log blocks in uneven size. The porcelain crafts, Chinese herbal medicine and varous types of mushrooms conveying the idea of health and medicated diet have been respectively placed in the wooden bookshelves and display cabinets. Modern elements like the balcony of Wine Culture filled with grape wine, the rattan dining table and interior impressionistic ink wash paintings have transmitted more multiplex aesthetic. Those ancient rhymes in different symbols which have quietly blended into the environment with the sense of modern crafts, and the newly developed oriental style is presented.

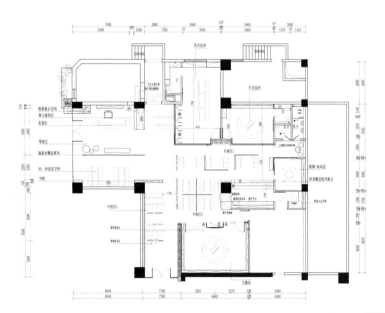

平面图

Plan

凯旋门七号会馆

7 Club Of Triumphal Arch

名称：

凯旋门七号会馆

设计公司：

河南鼎合建筑装饰设计工程有限公司

设计师：

孙华锋、刘世尧、孔仲迅

主要材料：

水纹砂岩、意大利木纹石、黑金花石材、黑钛金

Name:

7 Club of Triumphal Arch

Design Company:

Dinghe Architectural Decoration Design Project Co.LTD

Designers:

Sun Huafeng, Liu Shiyao, Kong Zhongxun

Major Materials:

Hydrological sandstone, Serpeggianto, Portopo stone, Black titanium

在九朝古都洛阳这样一个历史文化积淀深厚的城市，怎样将商业运营、客户体验与当下的设计潮流和厚重的地域文化巧妙地融合，给客户带来非凡的尊贵体验，是在设计初期定位时设计师需要考虑的重点。

For Luoyang, the capital of nine dynasties and the city with long history and ancient culture, how the designers skillfully combine the commercial operation, customer experience, current design trends with the local culture to provide marvelous and dignified experience for customers, has become a priority in preliminary orientation.

入口是客户体验的第一站，会馆外立面错落有致的石材分割巧妙地处理了建筑体量过大带来的沉重感，在带来强烈视觉冲击的同时也节约了造价。红黑相间的格栅与门套的结合强调了入口位置，也形成了畅怀迎宾的视觉效果。

Entrance is the first stop for the customer experience, as the well-arranged stone segmentation in the pavilion's exterior wall has well solved the heaviness due to the oversized mass of the building, bringing strong visual impact and also saving the cost. The red-and-black grid with the set door emphasized the entrance, and also bring in the visual perception.

进入会馆大厅，通道两侧的石材柱子、高悬的红色壁灯、通顶设计的中式花格、黑色描金漆柜等元素的有序排列形成了强烈的仪式感，凸显了客人的尊贵。

In the hall, the orderly layout of stone pillars, red lofty wall lamps, Chinese-style lattice, black cabinets with gold lacquer form the strong sense of ceremony and highlight the customers' distingnished position.

更让人震撼的是那两个挑高8米的中庭，尤其是正对入口的中庭，其上方璀璨的花瓣倾泻而下，汇聚成晶莹剔透的水晶牡丹，气宇轩昂却也柔情似水，体现出洛阳是牡丹之都的主题。从平面布局到装饰手法再到细节刻画，设计师举重若轻，使得厚重地域文化在空间中自然流淌，摒弃繁复的中式符号，保留中原文化的包容大气，融合现代人的审美情趣，给人带来一场不同文化碰撞的视觉盛宴。

There are two 8-meter atriums : The bright petals above the atrium directly facing the entrance are pouring down to converge into the crystal clear peonies, which are powerful but tender, highlighting Luoyang as "the City of Peony". From the layout to decoration method, and detail depiction, the designer dealt with them with ease to make the dignified local culture flow in the space naturally, abandening the complicated Chinese-style symbols, kept the essence of Central Plain Culture, and blended the modern people's aesthetic taste into the design to make a visual feast with the shock of various cultures.

陶俑
POTTERY FIGURINE

陶俑在古代雕塑艺术品中占有重要的位置。早在原始社会，人们就已经开始将泥捏的人体、动物等一起放入炉中与陶器一起烧制。到了战国时期，随着殉人制度的衰落，陶俑替代了殉人陪葬，秦始皇陵出土的七千兵马俑气势壮观，令人叹为观止。俑的使用是为了使死者能在冥世继续如生前一样的生活，所以俑真实负载了古代社会的各种信息，对研究古代的舆服制度、军阵排布、生活方式甚至中西文化交流都有着重要的意义。

大厅左侧的小中庭为藏宝阁休息区，高仿的唐代彩乐俑顽皮地立于两侧，高耸至顶的古董架下简洁的罗汉床可供人短暂休憩，整个空间都被浓浓的文化气息所包围却又不乏灵动通透。

The small atrium on the left side has been built as a treasure-house and resting area, while the highly-imitated tri-colored potteries of musicians in Tang Dynasty are standing on both sides, and the compact arhat bed under the towering whatnot is provided to the people to take a rest. The whole space has been surrounded by the deep cultural atmosphere.

一层除了风格各异的特色包间外还为散客或企业活动准备了一个面积达 240 平方米的散座区，满足各类客群需求的同时也让空间更加灵动和人性化。

Apart from the featured parlors in various styles, the ground floor also provides a 240-square-meter area with several randomly-placed seats for individual customers or corporate events to meet the needs for different customers and make the space much more flexible and humanized.

一层至二层交通方式由电梯和围绕藏宝阁的步梯组成，洞石铺就的地面将人们引向二层餐厅包间。

The means of transportation from the first floor to the second floor is formed by elevator and ladder surrounding the treasure-house, as the ground paved with the travertine can lead one into the parlors of the restaurants on the second floor.

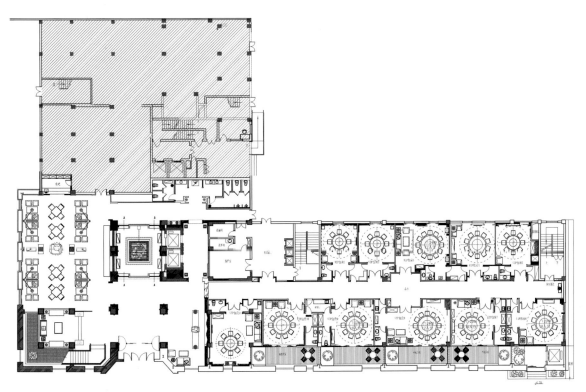

一层平面图
First Floor Plan

二层的包间中有几个特色包间，其中东宫（中式风格豪华包间）、西宫（新古典风格女性会所）和伊斯兰包间最具特色，呈现出尊贵大气、细腻典雅的空间氛围，满足了贵宾、高端女性客户及伊斯兰民族客人的个性需求，体现其人性化服务的宗旨。

There are several featured box rooms on the second floor, as the East Palace (Chinese-style luxurious box), West Palace (neoclassical-style female's club) and Islamic box are the most unique ones, which present the dignified, exquisite and elegant space, meet the characterized needs for guest reception, high-class female customers and Islamic guests, and embody the aim of people-oriented service.

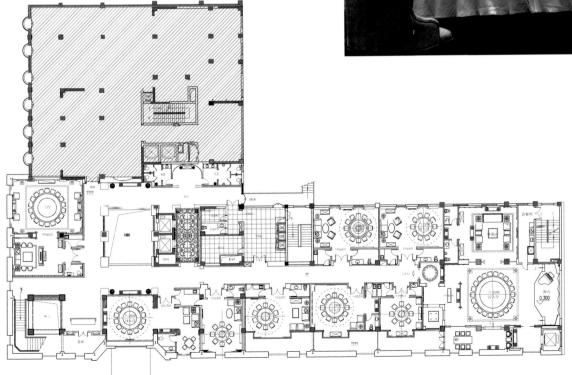

二层平面图
Second Floor Plan

苏园壹号

Suyuan No. 1

名称：

苏园壹号

设计公司：

河南鼎合建筑装饰设计工程有限公司

设计师：

孙华锋、刘世尧

摄影：

孙华锋

主要材料：

白木纹石材、古木纹石材、非洲紫檀、席编、硬包、城砖等

Name:

Suyuan No. 1

Design Company::

Dinghe Architectural Decoration Desigh Project Co.LTD

Designers:

Sun Huafeng, Liu Shiyao

Photographer:

Sun Huafeng

Major Materials:

Stone with white wood grain, Stone with antique wood grain, African padauk, Mat weaves, Hard rolls, City-wall bricks, etc.

对《牡丹亭》的大美印象成了此次苏园设计的主题。把声音与建筑，餐饮与文化很好地融合，并用高雅的昆曲艺术带动餐厅的发展，既提高了人们的生活品位，又推动了昆曲文化的发展。

The beautiful impression of Peony Pavilion has become the theme of Suyuan's design.The blend of sound and architecture, catering and culture has uplifted people's taste and boosted the development of Kunqu Opera.

原建筑为两层结构的现代风格售楼中心，我们有意识地将会所入口设计为通过庭院进入廊道再进入会所的格局，让客人有节奏地感受到中国园林的含蓄和精致。改造后的建筑为简约的徽派建筑形式。

The original buildings are a two story sales center with modern-style structure, as we have consciously designed the entrance of club into the courtyard through which one can get into the gallery and then get into the club, to allow the guests to have the cadent feeling of Chinese garden's connotation and delicacy. The transformed building is the minimalist Hui-style architectural form.

包间的设计以书架为依托体现出中国文人雅士的情怀，以大美、素雅、含蓄传递出会所的风雅气质。在陈设上，设计师采用西方新古典及新中式家具的混搭的方式，既创造了很好的舒适感又取得了极佳的视觉效果。

The design of restaurant parlors reflects the temperament of Chinese refined scholars, and transmits the grace and temperament with the great beauty, simplicity, elegance and connotation. In the display design, the mix of Western Neo Classical and new Chinese-style furniture is used to get not only good intimacy, but also terrific visual effects.

为了使厅堂版《牡丹亭》能在会所中演绎，设计师在兼做散台区的中庭采用了中国传统徽派建筑的榫卯木结构，并将地板做成升降地板以备演出、活动之用。宝蓝色布草的运用使大面积木色藤编的暖色得以平衡，布幔和竹帘的设置既让在散台区的客人有包间围合感，又打破了传统木构建筑的生硬。

《牡丹亭》
THE PEONY PAVILION

《牡丹亭》，是明朝剧作家汤显祖的代表作之一，共55出，描写杜丽娘和柳梦梅的爱情故事，是中国戏曲史上浪漫主义的杰作，与《紫钗记》《南柯记》《邯郸记》并称为"临川四梦"。此剧原名《还魂记》，创作于1598年。作品通过杜丽娘和柳梦梅生死离合的爱情故事，表现出追求个人幸福、呼唤个性解放、反对封建制度的浪漫主义理想，具有极高的历史价值和艺术价值。

To perform the spirit of "Peony Pavilion" in the club, we have used tenon-and-mortise timberwork of Chinese traditional Hui-style architecture in the atrium and designed the floor into the lifting platform for performance and activity. The use of royal-blue linen makes the warm color of large-sized rattan tone weaves balance well, as the drapery and bamboo curtain provides the guests in the public catering area the sense of enclosure and break the crudeness of traditional wooden architecture.

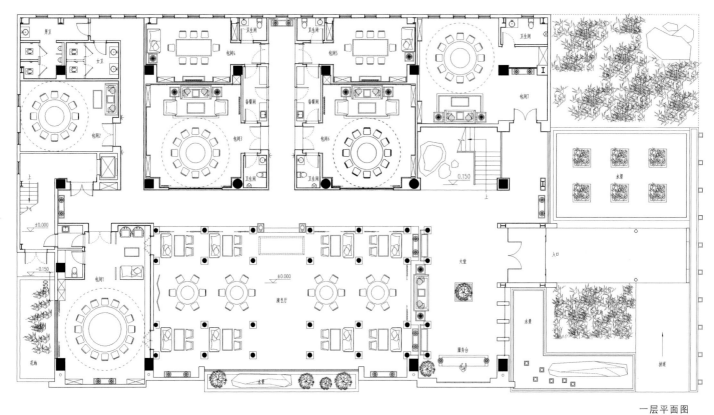

一层平面图
First Floor Plan

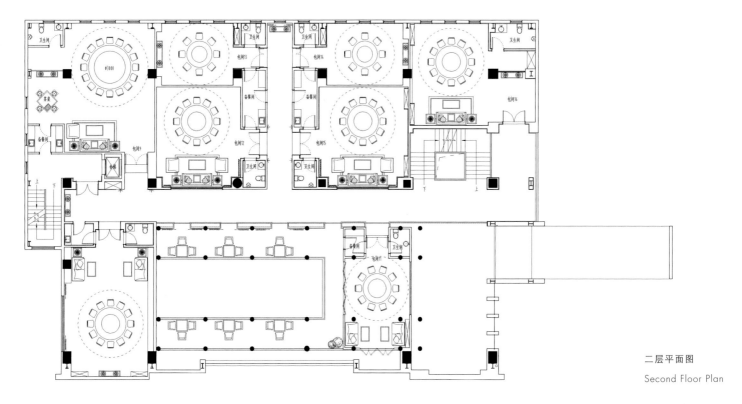

二层平面图
Second Floor Plan

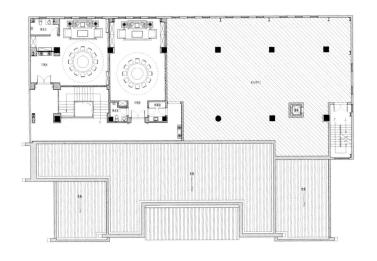

三层平面图

Third Floor Plan

屋顶平面图

Roof Floor Plan

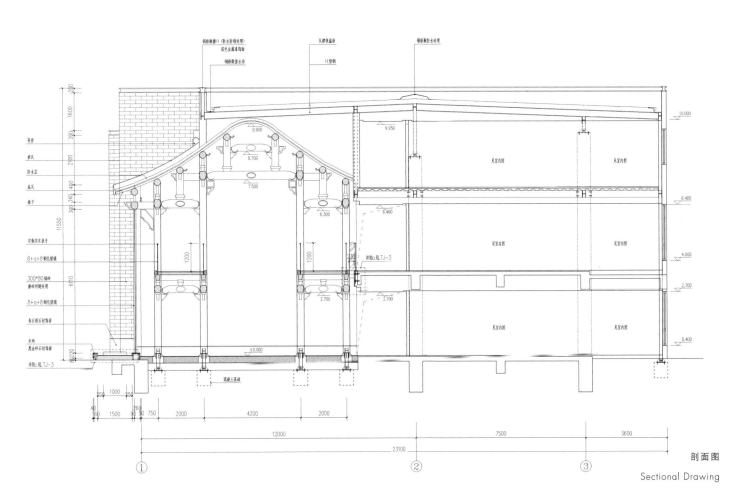

剖面图

Sectional Drawing

茶文化
Tea Culture

茶

中国是世界上最早发现和利用茶树的国家。唐代陆羽在《茶经》中说，茶是"发乎神农氏，闻于鲁国"。"神农尝百草，日遇七十二毒，得茶而解之"，所以说饮茶之始，是"食药同源"，食用在先，药用在后，饮用则是再后来的事。

中国最早饮茶史有所记载的是从西汉开始，而且是在四川（古巴蜀一带）。蜀人王褒所著《僮约》中提到"武阳（都）买茶"及"烹茶尽具"，说明在秦汉时期，四川产茶已初具规模，而且当时已经在生活中饮茶了，而且有专门的茶具。这也是历史上茶第一次作为商品出现，意义重大。江南初次饮茶的纪录始于三国，即是孙皓令韦"以茶代酒"的故事，当今在酒席上，在不能饮酒的情况下，人们还是延续着"以茶代酒"的习惯。

与 "禅茶一味"之渊源

茶真正的兴起始于唐代，因为唐朝是儒、道、佛三教合一，唐代禅宗因为是本土化的佛教文化，受到了文人雅士的推崇。和尚们不仅自己爱饮茶，在寺院还有专人负责茶事，另设有茶堂，不仅赠人饮茶还有"施茶"和"普茶"（僧俗同饮）等活动，使得许多信众都争相饮茶，促进了饮茶文化的传播。

禅宗和茶艺结合起来就有了著名的"禅茶一味"的说法。"禅茶一味"是佛教词汇，意指禅味与茶味是同一兴味。禅宗不讲苦修，讲究的是平常心，在平常中顿悟，遇茶吃茶，遇饭吃饭。正是有了禅宗寺院的大力推广，饮茶才逐渐在民间流行起来。被称为"茶圣"的唐代诗人陆羽所著《茶经》，是唐代茶文化形成的标志。此书概括了茶在自然和人文科学方面的双重内容，探讨了饮茶艺术，首次把"精神"二字贯穿于饮茶之中，把儒、道、佛三教融入饮茶中，强调茶人的品格和思想情操，把饮茶看作"精行俭德"，进行自我修养、陶冶情操的方法，首创中国茶道精神。

《萧翼赚兰亭图》，唐，阎立本。图中右为烹茶的老者和侍者，老者蹲坐在蒲团上，手持茶夹子正欲搅动刚投入釜中的茶末，侍童手持茶托茶盏，准备分茶入盏。高僧讲道的绘画主题，正好说明了在唐代茶与佛禅之间的密切关系。

《斗茶图》，唐，阎立本。反映了唐代市井中人们斗茶的场景。

凵 斗茶的兴起

宋代以前，中国的茶道以煎茶道为主。到了宋代，中国的茶道发生了变化，点茶法成为时尚。和唐代的煎茶法不同，点茶法是将茶叶末放在茶碗里，注入少量沸水调成糊状，然后再注入沸水，或者直接向茶碗中注入沸水，同时用茶筅搅动，茶末上浮，形成粥面。根据点茶法的特点，民间兴起了斗茶的风气。

斗茶，多为两人捉对"撕杀"，三斗二胜。决定胜负的标准有两条，一是汤色，即茶水的颜色。"茶色贵白"，"以青白胜黄白"。二是汤花泛起后，水痕出现的早晚。早者为负，晚者为胜。南宋开庆年间，斗茶的游戏漂洋过海传入了日本，逐渐变为当今日本风行的"茶道"。

《文会图》局部，宋，赵佶，台北博物馆藏，184.4 cm ×123.9 cm,绢本设色。赵佶即宋徽宗，是中国历史上最具文人情怀的皇帝，他非常喜欢设茶宴款待同是文人身份的臣子，并作画纪念之。从中我们看到宋代瓷器茶具的不同器型、颜色、摆放方式等，让人遥想当年的典雅情怀。

《文会图》局部，表现了宋代茶事的细节和点茶法。

而宋朝还有一种玩茶的游戏叫分茶，也称茶百戏。许多文人墨客都精于此道，认为是展现自己才学的绝佳机会。分茶就是在点茶的过程中，诵讨茶末和沸水相遇，变幻出各种奇异的画面来，花鸟鱼虫，四时风景等，最重要的是还要在这短暂的时间里面因应这些图像赋诗一首。

茶作为一种精神文化，是从品茗斗茶开始的。茶饮具有清新、雅逸的天然特性，能静心、宁神，有助于陶冶情操、去除杂念、修炼身心，这与提倡"清静、恬淡"的东方哲学思想很合拍，也符合佛教、道教和儒家的"内省修行"思想。因此我国历代社会名流、文人骚客、商贾官吏、佛道人士都以崇茶为荣，特别喜好在品茗中吟诗议事、调琴歌唱、弈棋作画，以追求高雅的享受。

匕　不同地域的茶事

我国大部分汉民族饮茶方法沿袭明清传统，以清饮雅赏的冲泡茶为主。但是中国地大物博，自然和人文环境差异较大，形成了几种具有代表性的饮茶方式。其中，潮汕功夫茶是清代以后的茶俗中最富特色的，还能体现唐宋以来的饮茶艺术余韵，而广式早茶则着重"吃"字，成都的茶则贵在那份休闲心境。

潮汕功夫茶起源于唐宋，兴盛于明清，后来更是影响到福建以及台湾的饮茶方式。功夫茶的"功夫"二字除了本领的意思，还指时间。潮汕功夫茶一般主客四人，主人亲自操作。首先点火煮水，并将茶叶放入冲罐中，待水开即冲入冲罐中之后盖沫。第一倒掉，雅称"润茶"，以初沏之茶浇冲杯子，目的在于造成茶的精神、气韵彻里彻外的气氛。再冲泡后，盖上壶盖，用沸水淋壶，提高壶内茶的温度，叫作"壶外追香"。斟茶时，四个茶杯并围一起，以冲罐巡回穿梭于四杯之间，直至每杯均达七分满。此时罐中之茶水亦应恰好斟完，余津还需"一点一抬头"地依次点入四杯之中，称为"关公巡城"和"韩信点兵"。四个杯中茶的量、色须均匀相同，方为上等功夫。最后，主人将斟毕的茶，双手依长幼次策奉于客前，先敬首席，然后左右佳宾，自己最末。

早晨上茶楼的生活习惯和社会风俗在广东由来已久，特别是在羊城广州，饮茶风气之盛，点心品种之多，堪称全国之冠。广东人上茶楼，一日三市，尤以早茶为盛。他们称早茶为"一盅两件"，就是一盅茶，加两道点心，广东人将此视为人生一乐。

四川的茶馆亦久负盛名，尤其在成都的大街小巷，茶馆比比皆是。"天下茶馆看四川，四川茶馆看成都"，成都人爱泡茶馆是出了名的，上万家大大小小的茶馆遍布大街小巷。四川茶馆中，人们饮用的是清一色的盖碗茶，充满了浓郁的地方特色。而北方地区由于不出产茶叶，茶馆里通常使用盖碗冲泡茉莉茶，同时还有各式点心、说书、唱戏等，这种大杂烩的形式，正是北方茶馆的一大特色。

匕　品茶的场所

茶席，是为了艺术地品茶而设置的场所，在茶席设计的过程中，从选茶、择水、备具，到茶桌台布、插花和字画等物品的摆设，以及茶艺师的服饰，都应该符合一个共同的要求，那就是把所泡之茶的特色最大限度地展现出来，让品茶的人得到最美好的感受。

茶席始于我国唐朝，诗僧与雅士对中国茶文化的悟道与升华，奠定了茶礼、茶道、茶艺的文化符号。至宋代，茶席不仅置于自然之中，宋人还把一些取型捉意于自然的艺术品设在茶席上，插花、焚香、挂画与茶一起称为"四艺"，常在各种茶席间出现。明代茶艺行家冯可宾的《茶笺·茶宜》中，对品茶提出了十三宜：无事、佳客、幽坐、吟咏、挥翰、徜徉、睡起、宿醒、清供、精舍、会心、赏览、文童，其中所说的"清供"和"精舍"，指的即是茶席的摆置。

以"枯木"为主题的茶席设计，选用质感粗糙的席面

茶席空间要根据茶的特性而设计，茶室布置得精美文雅，有书、画、插花等可供欣赏，此时的茶席就不仅仅局限在那一小部分，而是品茶的另一番天地了。茶席的布置主题先行，确定主题后，陆续选择相应的茶席元素。

席面设计的色调通常奠定了整个茶席的主基调，布置时常用到的有各类桌布，如布、丝、绸、缎、葛等，色彩和纹样多可选用中国艺术所崇尚的中性色调及中国传统纹样，质感上应重淳朴粗粝，避免轻浮。

竹草编织垫和布艺垫等兼有自然之美和工艺之美，而取法于自然的材料，如荷叶、沙石、落英铺垫等，则独辟蹊径，给人独特的审美感受；还有不加铺垫者，直接利用特殊台面自身的肌理，如原木台的拙趣、红木台的高贵、大理石台面的纹理等。

茶具，是整个茶席中的焦点，茶具的特色往往有启发主题的作用，温润淳朴的黑陶和紫砂与清新雅致的白瓷和青瓷，它们传递出的是不同的品茶心境。在现代审美的影响下，玻璃器皿、竹木茶具和铜锡茶具也成为了一种特别的选择。根据功能区分，成套的茶具要包括泡茶壶、饮茶杯、贮茶罐和辅助用具——茶则、茶夹、茶漏、茶勺、茶针、茶筒、茶炉、茶船、茶荷等）。

茶壶、茶杯。

贮茶罐

茶炉

选择什么植物以入茶境，关系着茶境文化的深层韵味。所以古代文人对茶境植物的选择极其严格，宜茶的植物只有竹、松两种。竹下品茶，是文人"无竹令人俗"的自然涌现；松下品茶，是雅士禅说情怀的迷人况味。

茶则

茶船

茶荷

茶席的插花，既可以是非常正式的插花作品，也可以是一两枝折枝花卉来做点缀，最重要的是花材的选择要符合茶席意境即可。

撰文 周航馨

107

古逸阁 · 茶会所

Guyi Pavilion-Tea House

名称：

古逸阁 · 茶会所

设计公司：

品川设计

设计师：

陈杰

摄影：

周跃东

撰文：

全剑清

Name:

Guyi Pavilion-Tea House

Design Company:

PURE' CHARM SPACE DESIGN ORGANIZATION

Designers:

Chen Jie

Photographer:

Zhou Yuedong

Article:

Quan Jianqing

茶会所的空间形态往往被束缚在某种意识形态中，但相同名称的空间种类的差异性则体现在设计师的匠心独运中。古逸阁·茶会所作为设计师"浮云"系列作品之一，"无像无相"是其对空间意境的全新诠释。这种创造的源动力让这个会所衍生出别样的气质，以木为媒，以禅为念，心清助道业，清苦得心定。

The spatial form of the tea house is often bonded by some kind of ideology, but the diversity of spaces with the same name shows the designer's ingenuity to their specific feature. As one of the works of "Floating Clouds" by the designer, Guyi Pavilion is a new interpertation of the space. Unique temperament of the tea house is derived from the driving force of creating. Wood as the material, Zen as the thought. Clear mind supports your practice, while austerity calm your heart.

茶会所虽地处繁华闹市，但设计师遵循"物尽其用是为俭"的理念，将一份古朴与清静浸润在空间之中，让目之所及的一切愈加耐人寻味。会所前的户外区域，地面用夯实的枕木铺就，周边的桌椅布置体现了木质、石质、竹质的交糅，透着一股自然苍劲的美。墙面的透明玻璃能呈现出会所内部的景致，它仿佛是一副取景框，涵盖的风景或许只是一个插着枯枝的陶罐，一把改良过的中式椅子，抑或是灯光留下的一抹影子。

Although the tea house is located in the downtown area, the designer has abided by the idea that "frugality is just to make fully use of everything", and put the plainness and peace into the space to make everything afford much more thought. The floor of the outdoor area in front of the tea house is paved by the rustic sleepers, while the

tables and chairs around intermix with the wood, stone and bamboo to provide a sense of natural and vigorous beauty. The transparent glass on the wall has shown the tea house's interior scene. It looks as if a viewing frame. The scene contained in it may be a terrine with a stick, a Chinese-style chair after improvement, or the shadows made by the light.

引导人们进入会所内部的地板是从附近老房子拆迁得来的旧木，凹凸不平的纹理自成风景，而那些或深或浅的不同色泽仿佛在默守一段尘封的往事，既给人留下遐想的空间，同时也衬托出这屋子的素雅氛围。内部空间的墙面也使用了这些旧木，它们带着一股时过境迁的味道，骨子里的文化归属让设计更加赏心悦目，让我们更容易找到共鸣。此外，会所内的柜体、台面、搁物架、门窗均以旧木作为设计的载体。或许它们已破旧、已衰败，但设计师挖掘出它们所蕴藏的蓬勃生命力，并将之运用在茶会所中，使他心中对传统文化的理解落到了实处。

What led people into the tea house is the wooden floorboards on the ground, which are old logs obtained from nearby demolished old houses. Their rugged textures have become a view, which provides much imagination to the customers and some sense of rustic elegance to the house. The interior wall coverings have also applied such old logs, making the whole

interior harmonious. Other than these, the cabinets, table-tops, shelves, doors and windows have all been re-designed on such materials. These materials may be old and decadent, but the designer dug out their hidden vitality and applied it in the design of the tea house, which is also a realization of his understandings of traditional culture.

除了旧木饰墙，天然麻布也是空间中重要的装饰材料，"素而不俗"是麻布的特色。前台区域的顶上悬挂着若干浮云状照明灯具，它们在光影的烘托下变幻出和谐的律动，并实现了空间氛围的蜕变。前台区域的背后是一个包厢，古朴自然的材质在其间和谐共处着。这些物件模糊了时间的概念，颇有一番自在的个性。

Apart from the old-wood-adorned wall, natural linen is an important material in the space, which is simple but original. The lightings in floating-cloud shape hung on the top of front desk area are creating the harmonious rhythm with the foil of light and shadow. There is a box room at the back of the front desk area. Rustic and natural materials are in harmonious coexistence. These have obscured the concept of time, but with some carefree personality remained.

在与功能区域的衔接上，"无像无相"是设计师追求的意境。一面灰砖砌成的墙，以方格作为透视，并点缀上烛光，让诗意在这古朴的空间中悠悠不尽，没有了浮躁，不见了仓皇。一旁的走道用青石做踏步，周边配以水景与地灯。走道的尽头是一面由石砾组成的墙，下面的石墩上放置着若干松果，寓意菩提。其上方用白色枯枝装饰，在灯光的映衬下显得张力十足。

A wall of grey bricks uses the squares as perspective, and embellishes with candlelights to make this space. The aisle uses bluestone as steps, with waterscape and floor-lights around it. The end of aisle is a wall bricked with gravel, while some pinecones are put on the rock-mound implying the existence of Bodhi. Some white deadwood are adorned on its above, full of dramatic feeling under the light.

菩提
Bodhi

菩提，佛教术语，是梵文"Bodhi"的音译，意思是觉悟、智慧，用以指人忽如睡醒，豁然开悟，突入彻悟途径，顿悟真理，达到超凡脱俗的境界等。释迦牟尼正是因为成就这种觉悟而成正果，世称佛陀。遵从佛陀的经教可修成菩提，按部派佛教的说法即成为阿罗汉，按大乘佛教的说法即成为佛，所以修证菩提是佛教徒的崇高理念。

与常规的佛像布置不同，设计师在此区域以意代形，将禅意与生俱来的气质展现得浑然天成。走道尽头的左侧是一个独立的品茶区，旧木、老物件依然是这里的主角，与传统文化一脉相承的灵动也愈加凸显。右侧是一个展示区，茶品、建盏、紫砂壶等物件陈列其中，它们是一种关于古朴与情境的东西，留下的是经过沉淀的生活。

Different from the regular arrangement of Buddha figurine, the designer has used the sense to replace the form in this area, and present the inherent temperament of Zen. There is an independent tea-tasting area on the left side at the end of aisle, as the old wood, old objects are still playing the leading role here, and the successive spirituality with traditional culture becomes much more charming. In a display area on the right side, there are

some objects such as tea products, black glazed porcelain, dark-red enameled potteries, are something about plainness and time, left by experienced life.

人若谢物，物未必不知。以物为善而无贵贱新旧之分，这是种人生的选择，也是种设计的态度。善用旧物成为设计师解决问题的一种态度，并由此衍生出一种新的生活方式。于是在古逸阁的空间设计中，材质之间的呼应与衬托、线条之间的交织与平衡、几何形态之间的构成与对比没有多余的一笔，不带丝毫拖沓。在这样没有喧哗的交流中，整个空间似乎变得香醇，人们的心情也变得明朗了起来。

When we express our thanks to the objects, it is very likely that they could sense it. To treat the objects well regardless of their states is a choice of life and an attitude towards design. Making good use of the old objects has become the designer's attitude to solving problems, which slowly derives a new lifestyle. Therefore in the design of Guyi Pavilion, there are various materials, lines and shapes mixing together and forming a beautiful balance without any redundance. Under such quiet communication between materials and texture, the whole space becomes a location with style.

❀ 盆景
Bonsai

盆景艺术源于中国，以后传到日本、越南和朝鲜，成为一种特有的传统艺术，约有一千二百多年历史。盆景以植物、山石、土、水等为材料，一般有树桩盆景和山水盆景两大类。盆景是由景、盆、几（架）三个要素组成的，是呈现于盆器中的风景或园林花木景观的艺术缩制品，具有缩龙成寸、小中见大的艺术效果，常被誉为"无声的诗，立体的画"。

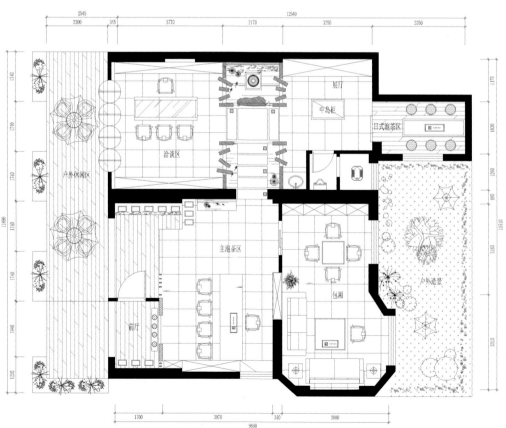

平面图

Plan

茗仕汇·茶会所

Celebrity Tea Club

名称:

茗仕汇 茶会所

设计公司:

品川设计

设计师:

陈杰

撰文:

全剑清

Name:

Celebrity Tea Club

Design Company:

PURE' CHARM SPACE DESIGN ORGANIZATION

Designer:

Chen Jie

Article:

Quan Jianqing

推开此扇方圆和谐的门扉，将喧嚣与浮躁都留给两旁的青葱翠竹，你我只管前往这淡然的所在，去览那久违的轻松。

When you enter Celebrity Tea Club, the first thing comes into your eyes are the green leaves of plants and bamboos on both sides of the path, which erases all noise and depress and makes you enjoy the relaxing moments that have long gone.

走过青石铺就的路面，苍劲的书法在轻薄的绢面上翻飞，光透过以此为屏风的隔断，令这些墨迹泛着久远的意蕴，散发出摄人心魄、无法抗拒的美。

Walking on the bluestone path, you are hard to miss the vigorous calligraphy on the translucid screen where these inks with in-depth meanings become more attractive than ever.

一具石像静伫在小径尽头，仿佛
已等了许久，而此处更像极了你
我最终的归宿，心的故乡之所在。

A stone figure stands still at the
end of the alley, as if waiting
for so long. The view here is
designed in a way that it is like
the place where our hearts
belong.

青砖墙上生长的绿植生机勃勃，
颇有一丝耐人寻味的寓意。

The greenery on the bluestone
wall is so vibrant as if they have
some implications.

吊顶之上奇异的灯如同朦胧的月先，只是换了修长的身姿，将来者的思绪拉得绵长。

Exquisite lamps on the ceiling looks like the vague moon in the sky but with a slimmer shape, which makes the viewers' thoughts long and lasting.

白色的鹅卵石洒落在路边，交错伫立的原木装饰板像把折扇慢慢推开，将空间的意蕴荡漾开去。入目尽是简约、古雅的明式隔断和家具，令古意弥漫在空间每个角落。

White pebble are spreading on both sides of the path, and all the crossing or paralleling facades form a folding fan which is slowly unfolding and spreading over the space. There are also many simple yet stylish Ming Dynasty partitions and furniture, which fill each and every corner of the space with ancient charm.

那一把古琴弹奏的又是哪一曲意味深长的古调呢？闲坐于此，喝一杯清茶，听一首古曲，可以忘了那流水般溜走的时光，想必这也是让设计师倍感满足的事情。

Which deep and meaningful ancient tunes are playing on this ancient instrument? Sitting on the chair, you sip a cup of tea and listen to an ancient melody, and all the time passing by would be forgotten in an instant. This must be a most satisfying moment for the designer.

❀ 撞水撞粉
TECHNIQUE OF ADDING WATER OR POWDER TO STILL-DRYING PIGMENT

由近代岭南绘画史上具有里程碑式意义的画家居廉、居巢，在继恽寿平以来的没骨画技法的基础上，创造性地总结出的一种绘画技法。此技法具体是指水与色、水与墨、色与色、色与墨之间的趁湿撞染，由此产生出自然渗化的效果。

拱形的青瓦垒成的隔断，形成一种独特的美感，光透过瓦片之间的间隔照射进来，配合着墙面的纹理，明暗之间好像藏着又一个天地。枯木之上一小盆生机勃勃的植物，蕴含了"枯木亦逢春"的哲思。

The partitions formed by the arching blue tiles constitute an exquisite beauty. The light shines through the gaps between the tiles, as if a whole new different space hidden between those lights and shades. With the texture of the wall coverings, the small pot of plant over the deadwood also implies the Eastern philosophical thought that the deadwood may still sprout again in spring.

✿ 瓦
Tile

或称屋瓦、瓦片。为亚欧常见的屋顶铺装建筑材料，具有防止雨水渗漏至屋内、隔热保温的功用。以粘土为主要原料，经泥料处理、成型、干燥和焙烧而制成。瓦在中国西周初期即已产生。而希腊奥林匹亚兴建的赫拉神庙则是最古老的有屋瓦的西方建筑。

简洁的线条，给予空间纯粹的力度与美学，可谓实用性与美学的完美结合。而那满墙的各式雕花，令人为之心动，想必是花了无数时光和心思收集而成的，怎能不令人为之折服。

Simple lines applied here provide strength and beauty to this space, which is a perfect combination of practicality and aesthetics in design. Various carvings on the wall are so charming that they must take the design team myriad of years to make and collect.

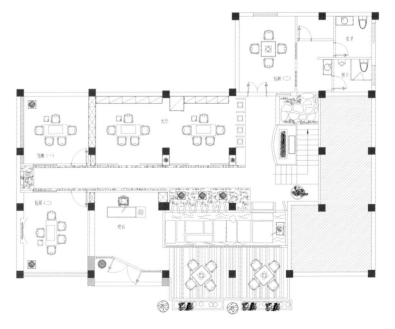

平面图
Plan

茶禅品味

TEA, ZEN AND ITS TASTE

名称：

茶禅品味

设计公司：

唐玛国际设计

设计师：

施旭东

摄影：

周跃东

主要材料：

火烧石、青石板、金刚板、文化石、仿古砖、槽钢、黑镜

Name:

TEA, ZEN AND ITS TASTE

Design Company:

TOMA

Designers:

Shi Xudong

Photographer:

Zhou Yuedong

Major Materials:

Burnt stone, Blue flagstone, Diamond crown, Archaized brick, U-steel, Black mirror

这是位于世贸外滩的敬茶坊。茶馆的一楼大厅部分空间为挑空式设计，一个方形玻璃吊灯如同一个透明小屋悬于接待台上方，极为别致，并且在上下映射之间从视觉上扩大和拔高了空间的结构。而一旁的佛像、灯笼及窗边的造景等独具中式风格的软装饰品，也一并被设计师规划到此处，成为辅助空间的点睛之笔。

The TeaHouse is located near Shanghai Mart and the Bund. The designer has designed a unique architectural structure floating above the main hall of the tea house on the first floor. A square box of glass pendant light looks as if a transparent house floating above the reception desk, which creates a spectacular scene, visually enlarging and elevating the structure of the space. While some Chinese style furnishings, such as the Buddha figure, lantern and scenery by the window, are also laid out in this spot, which become the highlights of the design.

茶馆的玄关处是一面古朴的石砖以不规则的拼贴方式拼贴成的背景墙，与门边漆成白色的树枝、泛着暖黄色光芒的灯台相映成趣，刹那间便抓住了人们的视线。进厅左侧的品茶区，设计师用一块块青砖错位叠放，垒出三堵镂空的隔墙，隔出三间茶宅，因方位错落，形成隔而不断的独特"包厢"，特殊的处理手法让人们虽处同一空间，却有"此处"与"彼处"的感受。

In the hallway of the tea house, there is a background wall covered with irregular primitive stone tiles, which constitutes an interesting scene and capture people's attention, together with the branches painted in white and the lamps with warm lighting. The designer used the bluestones to build up three cut-out partition walls, which skillfully create three individual yet connected box rooms. This unique partition technique forms people's concepts of "here" and "there" within the same space.

茶馆主体区域的地面铺以青石板与仿古砖，粗犷的自然肌理显得厚实而流畅，仿佛刻满了时间的痕迹。为了中和这些材质的硬，设计师在细节处，用了行云流水的"茶"字、藤编的灯笼，以及装饰玻璃上的祥云图案。那些做工精良的仿古家具、精美绝伦的青花瓷器等细微之处的累积让空间显得更为饱满。

The ground on the main area of teahouse has been paved with the blue flagstone and rustic bricks, of which the rough natural patterns are solid and fluent, as if having been instiled with trace of time. In order to alleviate these tough textures, the designer concentrates on details: the beautiful character of " 茶 (Tea)", rattan weaved lanterns, auspicious clouds patterns on the ornamental glass, and those exquisite antique furniture, blue-and-white porcelains, all made the space abundant.

祥云图案
AUSPICIOUS PATTERN OF CLOUD

"祥云"图案来源于我国古代的云纹。古人出于对云的敬畏，在纹饰上变化出各种和云有关的图案，运用于器物、服饰、建筑等方方面面，特别是在宗教中大量使用。同时云纹样也有着很多美好的寓意，表现了人们对万事万物希冀祝福的心理意愿和生活追求。而在 2008 年北京奥运会上，"祥云"纹样也被广泛使用，特别是火炬的创意灵感，来自"渊源共生，和谐共融"的"祥云"图案，给世界各国的人们留下了深刻的印象。

走进静茶访，偌大的空间在设计师精心规划下，既装出了大气磅礴之感又不乏细节的推敲，这种繁花落尽的寂静之美，来自于东方传统文化中"禅"的思维方式，包含了丰富的精神内涵，成就了空间的美感，体现了静茶访的设计精髓。

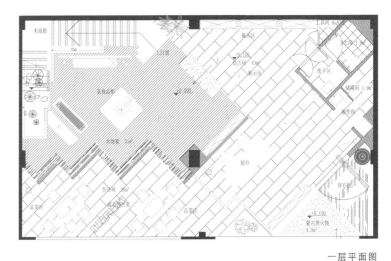

一层平面图
First Flood Plan

In the tea house, the large space is filled with imagination from the designer's elaborate planning. With the grand and magnificent decorative style, precise and meticulous structure, the quiet beauty of subtlety like the fallen flowers which comes from Zen philosophy in the traditional oriental culture, including rich spiritual intension.

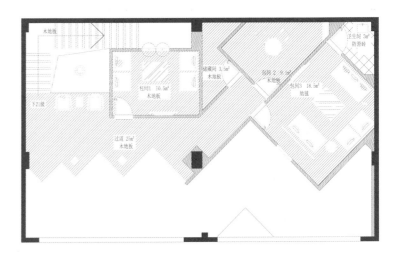

二层平面图
Second Flood Plan

熹茗会

XI MING HUI

名称:

熹茗会

设计公司:

唐玛国际设计

设计师:

王家飞

摄影:

周跃东

主要材料:

仿古木地板、硅藻泥、青石

Name:

XI MING HUI

Design Company:

TOMA

Designers:

Wang Jiafei

Photographer:

Zhou Yuedong

Major Materials:

Antique wood flooring, Diatom ooze, Bluestone

这是一个高端茶叶会所空间。设计师以江南小镇作为设计的灵感来源，以室内剧的表现方式，融合现代的美学来诉说一个小镇的故事。石板青青、小巷弯弯、丝竹悠扬、茶韵飘香……设计师通过塑造这样一系列的意境来表达对中国传统文化的尊重与敬仰。

The space is a high-end tea club. The designer was inspired by the small towns in the south of Yangtze River and decided to design the space applying the technique of indoor drama and blending in modern aesthetics. The blue stone flooring, the winding alley, beautiful flute music, the fragrance of tea...the designer creates such scenes and landscape to express his respect and admiration towards Chinese traditional culture.

水，既是中华文化的表现载体，又是各个空间相互呼应的有机元素，贯穿整个空间的水带低调内敛，让人联想到中国文人的气质。在简约硬朗的空间中，以中式家具和工艺品做点缀，呈现出庄重与优雅的双重气质。

Water, is not only the representation of Chinese spirit, but also the organic element through which every space works in convert with others. The water elements that run through the whole space are quiet and reserved, which reminds us of the temperament of traditional Chinese scholars. Chinese style furniture and art crafts embellished the minimalist space, expressing both solemn and elegance at the same time.

❁ 辅首
FUSHOU (CHINESE TRADITIONAL DOORKNOB BASE)

辅首是安装在大门上衔门环的一种底座，是中国传统的大门装饰。中国古代文献中对铺首的记载基本上为门上所用，例如《说文解字》上说："辅首，附着门上用以衔环者。"中国在春秋时代便已使用辅首，至今逾 2 000 年。辅首反映历史的变迁，可显示使用者的地位。

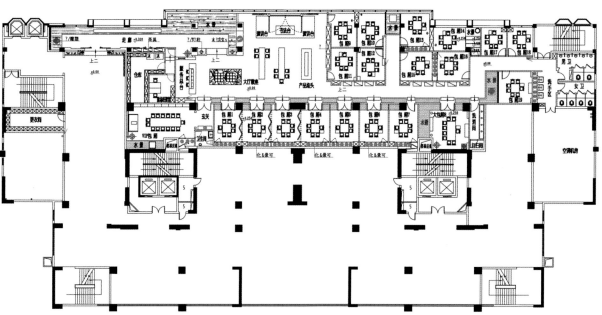

平面图
Plan

唐会

CLUB TANG

名称：

唐会

设计公司：

唐玛国际设计

策划：

施旭东

设计师：

洪斌，陈明晨，林民，胡建国，王家飞

摄影：

周跃东

撰文：

全剑清

主要材料：

原木做旧、艺术漆定制、腐蚀耐候钢、亚麻、钢丝网

Name:

CLUB TANG

Design Company:

TOMA

Planner:

Shi Xudong

Designers:

Hong Bin, Chen Mingchen, Lin Min, Hu Jianguo, Wang Jiafei

Photographer:

Zhou Yuedong

Article:

Quan Jianqing

Major Materials:

Antique-finished log, Customized art lacquer, Weathering resistant steel, Linen, Steel wire gauze

在唐会的入口处，略显窄小的门框似乎隐匿在周边的竹林之中，一抹灯光伴着竹叶沙沙的轻响，搭配上拴马石和传统的人物雕塑，建筑的诗意便悄然而生。

In the entrance area of Tang Hui, the narrow door frame seems to be hidden among the surrounding bamboos. The light, the sounds of bamboo leaves, the hitching stone, and traditional figure sculpture, all these elements constitute the beautiful poetry of architecture.

白色的地面与周遭的古朴形成强烈的视觉反差；叠石搭成的水幕墙，清水顺流而下，水声潺潺。白色的洗手台上方悬置着一根水管，当有人走近时，水便能自动流出，这种人性化的细节设计，提升了整个会所的空间品质。

The white flooring forms strong visual contrast with its rustic surrounding. The waterscape wall is piled up by stones, and the water flow down like waterfall, which creates beautiful sounds to the space. There is a water tube hanging above the white wash basin, and whenever anyone walks near it, the water may automatically flush out. These design details greatly elevate the style of the whole space.

❋ 竹
BAMBOO

竹子的发源地是中国，竹子的文化历史在中国非常悠久，被国外友人称为"竹子文明"。竹子代表着高贵、典雅、纯洁、谦虚，是一种精神文明的象征。竹子即是四君子之一，更位列岁寒三友之首。文人名士颂竹、画竹，并不单单是歌颂竹子的形态美和意境美，最重要的是歌颂竹子"宁折不弯"的品格和"中通外直"的度量。

吊顶以及部分墙面用不同纹理的中式雕花构件来装饰，每一块雕花板都有其独特的美，大小不同、花纹各异却又能整齐排列，在变化中展现出一种秩序的大气之美。

The ceiling and some parts of the wall are embellished by Chinese carving patterns and each carved board has its own unique beauty.

钢制的楼梯与一旁的毛石地面均带着硬朗的特质，而红色的木质扶手则给这个古朴沉静的空间，注入了一丝柔情与活力。楼梯口所面对的墙面依然用做旧的钢板来装饰，钢板上面是设计师亲手描绘的荷花图案。黑色在传统文化中代表着"水"，而这种围合式的布局也是对传统四合院的新诠释。设计师潜移默化地将传统气息沁入到会所设计中，身在其中，浮躁的心绪也在不知不觉中静了下来。

The steel staircase and the rugged flooring both have the quality of roughness, while the red wooden handrail instills this tranquil space with some gentleness and vitality. The wall against the stairway entrance is furnished with rustic steel, on which there are many lotus patterns hand-painted by the designer. The color of black represents water in traditional culture, whereas such enclosure layout is also a new interpretation of traditional Chinese quadrangle dwellings. The designer has quietly instilled the traditional taste into the space, where people would become tranquil unconsciously among this atmosphere.

楼梯区域，设计师用现代的几何解构思想来表达一种文化的碰撞与融合，并让空间在轻与重之间和谐共融。楼梯的下面铺上了白色的沙石，缀以风灯、铜壶等物件，颇有一番悠然自在的意境。

The designer has expressed the collision and fusion among cultures with the modern thought of geometric deconstruction, and made the space more harmonious between lightness and heaviness.The white gravel is paved below the stairway, studded with objects like kerosene lamp and copper pots. They have blurred the concept of time, but with some carefree atmosphere.

在二楼空间的墙绘上，三坊七巷建筑的写意与西方油画的写实搭配在一起，并在画的尽头虚拟上江水、帆船、海鸥等景致，让人不自觉地徜徉在艺术的氛围里。在会所的包厢中，暖色的麻布铺陈在一侧墙面上，自然朴实的材质和纹理让人感觉更加贴近自然。另一侧的墙面上则描绘了一幅古代文人的画像，绵延其中的人文精神削弱了元素间的冲突，适度差异让空间更加灵动。

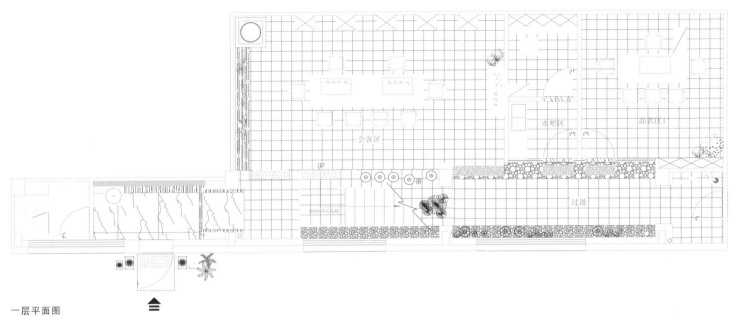

一层平面图

First Floor Plan

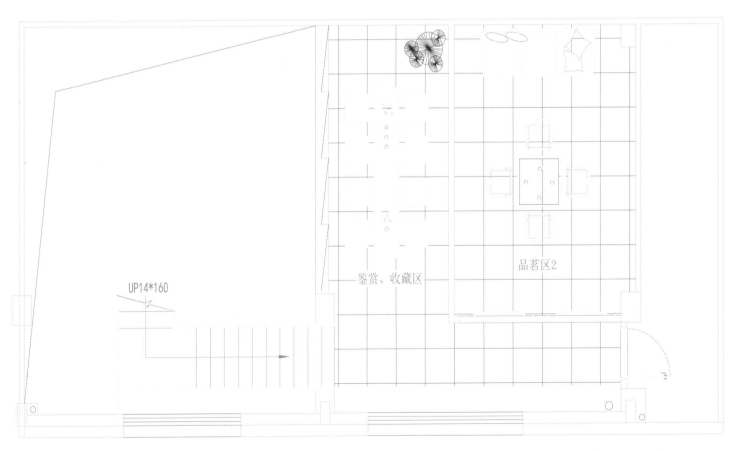

二层平面图

Second Floor Plan

On the second floor, the architecture in the Three Lanes and Seven Alleys has been matched with the realism of western oil painting, and the scenery like river, sailing vessels and seagulls immerses us in the art. In the club's balconies, the warm-colored linen is elaborated on the wall, with refreshing, natural and plain texture.

The image of ancient literati is painted on the wall of another side, as the continuous humanistic spirit has weakened the conflict among the elements to make their balanced difference enlighten the space.

静茶访·西湖店

Jing Tea House, West Lake

名称:

静茶访·西湖店

设计公司:

道和设计机构

设计师:

高雄

摄影:

周跃东

主要材料:

黑钛、水曲柳、黑镜、毛石、铁架花格、木纹砖、蒙古黑火烧板、墙纸白色烤漆玻璃

Name:

Jing Tea House · West Lake

Design Compeny:

DAOHE DESIGN

Designers:

Gao Xiong

Photographer:

Zhou Yuedong

Major Materials:

Black steel, Manchurian ash, black mirror, rubble, iron shelf and lattice, wood pattern tile, Mongolian black fire plate, white lacquered glass

本案是静茶的第三家店，也是静茶的形象店，位于美丽的西湖湖畔，为了更好地表达出静美意境，设计师利用了建筑的结构，结合中国山水的意境，分割出两处水景，使整个设计时时流露出中国传统的诗意和美学意境。

This project is the third division of Jing Tea House. Locating at the bank of the beautiful West Lake, the tea house also acts as an image store for the brand. In order to express the feeling of tranquility and elegance, the designer makes full use of the architectural structure by dividing the space into two sections of water landscapes and combining the traditional artistic concept of Chinese landscape into the space.

以中式窗格为元素的铁艺花格，从门头一直延伸至店内前台凹凸错落的石材背景墙，大大小小的"飞碟灯"从吊顶上高低错落的悬垂下来。白色大理石前台是大面积黑色中的点缀，大理石的分割设计很好地缓解了收银、咨询等功能给人的直观印象，与整个环境更加协调。在背景墙后面，是用格栅门分开的包厢，古韵古香的鸟笼灯更多强调抒发设计师的主观情趣，让观者臆想出一个真正的宁静致远之美好之地，这才是最自然的一种形式美感的东西。也符合了静茶其静之意，设计者更结合这个韵意，将之融于作品中，所谓静茶静美，君子淡泊。这正是设计师所理解的意境。

Large scale of iron lattices with Chinese elements are being used in the design, from the front door extending all the way to the embossing stone walls at the reception area. And various sizes of pendant lamps in unique UFO shape hang in the lattice, forming a scattered wave.White marble reception desk is a small ornament under the background of complete black. The dividing shape of the desk perfectly resolves the problem that some practical functions, such as cashier and inquiry, may break the overall atmosphere of the space.Behind the back walls, the space is diving into different compartments by the grilled doors. The brilliant designs of antiqued lamps in the shape of bird cages emphasize the designer's subjective taste and recreate a tranquil and beautiful Neverland for the viewers, which is the most natural way of expressing beauty in the form. On the other hand, the designer combines the meaning of tranquility into its design, which is also the symbol of the tea house.

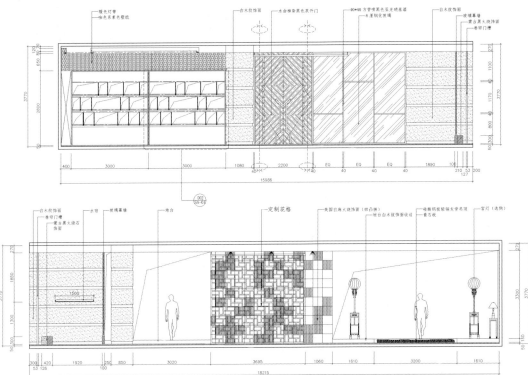

立面图

Elevation

✿ 官帽椅
CHINESE ANTIQUE ARMCHAIR

官帽椅为硬木家具，属扶手椅的一种，其定型可追溯于宋代，因搭脑处形似官员所戴纱帽上的帽翅而得名。从形制上来看，官帽椅大致可分为四出头官帽椅和南官帽椅两类。所谓四出头是指在椅子的搭脑两端及扶手前端处均延伸出头。南官帽椅的特点是在椅背立柱和搭脑相接处作出软圆角，由立柱作榫头，横梁作榫窝的烟袋锅式做法。

平面图
Plan

夷尊茶业

Yizun Tea

名称：

夷尊茶业

设计公司：

道和设计机构

设计师：

高雄

摄影：

李玲玉

主要材料：

黑钛、黑镜、墙纸、木纹砖、条纹砖、实木块、编织板、毛石

Name:

Yizun Tea

Design Firm:

DAOHE DESIGN

Designers:

Gao Xiong

Photographer:

Li Lingyu

Major Materials:

Black titanium, Black mirror, Wallpaper, Wood blocks, Streak bricks, Solid wood, Weaving board, Rubble

武夷岩茶品质独特，它未经窖花，茶汤却有浓郁的鲜花香，饮时甘馨可口，回味无穷。

Wuyi rock tea has its unique quality. The tea has rich fragrance of flower without scenting. It is so sweet and delicious with an endless savor.

"夷尊"用最浅显的文字记述武夷岩茶在众多茶叶种类中脱颖而出，崛起于岩壑之间，带给世人品饮艺术的一份清香。茶是国饮，茶香飘扬千年，我们对茶的热爱，是源于对生命本体和大自然的认知与感悟。

"Yizun" is the least word to describe how tea could stand out in various types, and bring people a fragrance of drinking art. Tea is the national drink with more than one-thousand-year history.

✽ 窨花
TEA SCENTING

窨（音义同"熏"）花又称窨制花茶，制造花茶的主要作业，即在烘焙茶的过程中，加入花一同窨香，让茶吸收花的香气而成。东方饮用花茶的记录最早可追溯至中国唐代，不过直到南宋时才有了极少篇幅关于茉莉花窨茶的记录。到了清朝咸丰年间，福建福州成为当时中国窨制花茶的生产基地，花茶开始大规模地在工厂里生产烘制，其中以茉莉花茶为大宗。

墙面起起伏伏的陶质马赛克令人联想起砖与火做茶的古老工艺，在灯光下呈现出斑斑驳驳的光影效果。与背景墙为突出LOGO而设置的射灯不同，饮茶区的灯光十分柔和，桌面凹凸不平的质感吸引人们伸手触摸，带给人轻松舒适的体验，设计带来的享受也从视觉延伸到触觉。

The embossed pottery mosaics remind us of the ancient technique of tea-making with bricks and fire. The lamps in this tea-sipping area create gentle effect of light and shadow, which is different from the spotlights set for the logo. The embossed table tops attract people to lay hand on them, which bring in comfortable experience. The sensation created by the design here extends from vision to the sense of touch.

因着对茶的热爱，设计茶店成为一件轻松的事儿，用心品味，凝精聚神，由茶的实体抽离出意象，并且让这些"意象"成形，在这方面设计师与业主达成共识——让设计和茶变得简单、自然。

To design a tea shop can be an easy thing for someone who loves tea. With the love for tea, one can taste the fragrance by heart, and focus on pulling the images away from the solid to let these "images" forming. In this respect, we have reached an agreement with the owners, to make the design and tea simpler and more natural.

平面图
Plan

禅
Zen

禅

📅 禅宗的起源

"禅"在古代东方和现代世界的许多地区极为流行。禅宗之前，一般所说的"禅"主要是指以"四禅八定"为主要内容的禅定，这也是"大乘六度"（布施、持戒、忍辱、精进、禅定、般若）之一。

禅宗，则是释迦牟尼佛教心法，与中国文化精神结合，形成中国佛教融合古印度佛教哲学最精粹的宗派。在佛学中，"禅定"是大小乘共通的行持修证的方法，"禅定"的原名为"禅那"，又有中文的翻译为"静虑"，后来取用"禅"的梵文原音，加上一个译意的"定"字，便成为了中国佛教惯用的"禅定"。

简单来说，禅主要是人的一种精神修持方法，是信奉者一种体悟真理或最高实在的方法，是其摆脱外界干扰、保持内心平静的方法。通过冥思修行寻求个人顿悟是禅宗的重要部分。相传中国禅宗初祖——6世纪印度僧人菩提达摩，曾经面壁多年，扯下自己的眼皮以防睡着，表明禅宗所倡导的个人戒律强度。

禅宗素来被称为教外别传（在如来言教以外的特别传授，指不依赖佛经，而靠自身感悟来体会佛理）的法门，历来相传，释迦在灵山上，对着百万人天，默然不语，只轻轻手揲一支花，向大众环视一转，大家都不明白他的寓意，只有大弟子摩坷迦叶，会心一笑，于是释迦则宣布："吾有正法眼藏，涅槃妙心，实相无相，微妙法门，不立文字，教外别传，付嘱摩坷迦叶。"这就是禅宗的开始，阿难为第二代祖师，历代相传，到第二十八代菩提达摩大师东渡而来我国，将禅宗文化带到中国。佛教的禅思想传入中国后，一开始并没有立即形成一个独立的佛教宗派。禅宗形成后，中国的禅思想则主要表现为禅宗的发展。

禅宗在中国唐代早期充满了渐悟和顿悟之间的争论。六祖慧能主张"顿悟"影响华南宗派，称为"南宗"；慧能的同门师兄神秀主张"渐悟"被称为"北宗"。到了唐朝晚期，顿悟派成为主流，强调自性，直指人心、见性成佛，师徒间直接传承。禅宗故事多侧重在"不立名相"的一面，禅宗祖师、禅僧或是白丁、或是帮厨，但自幼就极富慧根，获得彻底的"顿悟"。禅宗与弟子间的对话，强调了矛盾和不受束缚，明显受到了道家哲学文化的影响。

屮 在日本的传播

禅宗在日本成立较早，于日本镰仓时代（1192—1333）从印度经中国传到日本，并形成了日本的临济宗、曹洞宗、黄檗宗。

禅宗从中国传入日本后，逐步实现了从中国禅到日本禅的演变，并同步实现了从禅的贵族化到世俗化的转变。室町时期，禅宗在日本的发展初步呈现出一种世俗化倾向。此后，随着枯山水庭园的盛行、茶庭的普及等，禅在日本最终完成了贵族化到世俗化的转变过程。

禅宗强调极强的自制力，并引发了茶道，以及"遵循自然之道，崇尚自然主义"的艺术美学思想——可以说，禅是日本的灵魂，是日本文化的精髓。在日本，人们对于那些和禅一起兴盛起来的茶道、花道、绘画、建筑、武术等诸般技艺，往往习惯于到禅里面寻求其精神内蕴。例如，禅宗庭院内的树木、岩石、天空、土地等常常是寥寥数笔即蕴涵着极深寓意，在修行者眼里它们就是海洋、山脉、岛屿、瀑布，一沙一世界，兼具豪放与细腻之美。

枯山水庭院，用抽象的手法，在充分注意与自然景观相结合的同时，力图表达出一种和谐、洗练、枯寂、空无、清幽的禅之意趣。"园林中充满了日本特有的禅味和涩味"，充分地反映了禅宗文化对设计的影响和渗透

茶庭，在日本指与茶室相配的庭园，是日本庭院艺术中很有民族特色的作品种类。

屮 雅致的禅文化

尽管禅宗并不追求世俗以及传统，但是唐代的禅师被邀请到朝廷，解决了学说上的争论。而宋代时禅宗绘画达到了一个空前的高度，题材多数涉及古怪的禅僧、饱经苦难的悟道散圣、禅会图（常通过禅师和儒者间的对话回答来体现禅机）、禅宗祖师、佛以及菩萨。

《出山释迦图》，南宋，梁楷，主题直指释迦牟尼在采取不极端苦行的"中间道路"，前往深山禁欲苦修、寻求觉悟的岁月。梁楷在此画中塑造了一个没被理想化拔高的释迦，性情抑郁的释迦牟尼，独自一人忍受着寒冷的折磨。这种描绘佛祖的方式也许会使观者惊讶，但正和禅宗风格相称。

《六祖斫竹图》，南宋，梁楷，描绘了慧能在砍柴中获得了"悟"的场景。这种简率而含蓄的描绘方式和禅宗的态度一致，主张在在寻常事物以及事情中求得"道悟"，主张"悟道"前后看待世界即要不同也是一样。

禅宗强调佛教的修行活动不能脱离世间，这也是它的一个重要特色。用禅宗自己的话概括就是"佛法在世间"。中国以儒家为主要代表的传统文化通常要求人们积极参与社会生活，在社会生活中体验真理，特别是南宗系统，并不追求那种与世俗世界完全不同或毫无关联的涅槃境界或清净世界，而是认为不能离开世俗社会去追求涅槃境界。禅宗的"不离世间觉"就是强调修行不能脱离社会生活，要在现实世界中去追求对自身心中佛性的认知。

禅宗的这种特性，注定了它要与中国的世俗文化相融相生，派生出"农禅一味"、"禅茶一味"和"禅意空间"的雅致禅文化。"农禅一味"，也就是禅宗僧人自耕自食，农作和禅悟并行不悖，世出世间浑然合一的精神。许多羡禅向道的文人便以诗作描述农事来表现禅理禅意，称为"农禅诗"。

有"诗佛"之称的王维，其农禅诗大多属于"观农家而说禅"的诗。如《渭川田家》诗曰："斜光照墟落，穷巷牛羊归。野老念牧童，倚杖候荆扉。雉雏麦苗秀，蚕眠桑叶稀。田夫荷锄至，相见语依依。即此羡闲逸，怅然吟《式微》。"诗歌呈现的是一幅农家晚归图：夕阳西下，牧童归家，农夫返途相语，老人倚门相望。此诗核心意旨在一"归"字，回归自然，隐逸田园，是禅宗核心意旨之一。王维从农家晚归景象中体味到了闲适散逸的禅之意味，从而产生了归隐田园的意愿，故有此作。

禅宗文化还与茶文化融合在一起，产生"禅茶一味"的思想观念。禅茶文化学者陈云君曾这样说道："'禅'是一种境界。讲求的'禅茶一味'，'禅'是心悟，'茶'是物质的灵芽，'一味'就是心与茶、心与心的相通。"

茶与禅的相通之处在于追求精神境界的提纯以及升华。饮茶时我们需要平心静气慢慢品尝，参禅则要静心息虑体味，茶道和禅悟都看重主体感受，需要进行静心体会。禅是中国化的佛教，主张"顿悟"，而在饮茶的过程中得到精神寄托，也是一种"悟"，所以说饮茶可以得道，茶中有道，禅和茶就连接起来了。

中国文人对禅道的钟情，也融合在中国的园林和家居布置中，这种气质上的传承，在现代则演化为一种注重空灵气氛营造的独特格调——禅意空间，它主要是通过对空间的留白处理而实现的。空间中的留白未必就代表空无一物，它指的是设计师通过巧妙地运用视觉上"不在场"的部分，取得令人惊叹的整体设计效果。在注重禅意的中式空间中，设计师必须对细节十分敏感，能觉察出颜色、空间、光线、比例和设计目标之间的毫厘之差。

禅意中式的氛围或元素要能带给人们一种或明净闲适、或深远庄重的整体感觉。例如，我们不但可以通过结构、色彩和材质等实质性方面的简化，还可以通过有意无意之间的删繁就简，营造出明朗清晰的空间效果，并在此过程中获得一种洞烛幽微的力量。

撰文　周航馨

云音 · 禅会所

Yunyin Zen Club

名称：

云音 • 禅会所

设计公司：

河南鼎合建筑装饰设计工程有限公司

设计师：

孙华锋、刘世尧、孔仲迅

主要材料：

橡木、壁纸、生态木、黑镜、机刨石等

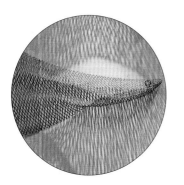

Name:

Yunyin Zen Club

Design Company:

Dinghe Architectural Decoration Desigh Project Co.LTD

Designers:

Sun Huafeng, Liu Shiyao, Kong Zhongxun

Major Materials:

Oak, Wallpaper, Greener wood. Black mirror, Planed stone, etc.

禅修是近年来兴起的一种新的修身方式，与瑜伽身体修炼不同，禅修更多地关注人精神方面的修炼，是通过诵经、坐禅、抄经、讲经的修禅方式让人达到放空内心，感悟并升华精神的一种修炼。本案以禅修为主线，兼容了禅茶、SPA、茶餐等功能形式。如何让空间自然相融又能动静相隔是设计对空间处理的要点，因此，在设计中三层主要分布了接待、禅茶、茶餐等功能。相对三层"动区"设置，二层更偏向"静"，SPA、禅房、抄经室的设置让客人更能够"静"心、安定。

Zen meditation is a new way of self-improvement arisen in recent years, it is different from the physical cultivation of Yoga. Zen meditation focuses more on the human's mental cultivation, which allows one empty his inner heart and sublime the spirit. The Project uses Zen meditation as the mainline containing the functions like Zen tea, SPA, tea meal, etc. How to make the spaces blend together and separate movement from peace is the design's priority to deal with the space. Therefore, functions like reception, Zen tea and tea meal have been mainly distributed in the third floor of the design. Opposite for the "movement area" in the third floor, the second floor is in much favor of "peace". The setting of spa, meditation room and room for copying the Buddhist classics makes the guests much more "peaceful" and stable.

禅修
BUDDHIST MEDITATION

修行禅定的行为被称为禅修。禅修，巴利文意思是"心灵的培育"，就是把心灵中的良好状态培育出来。在佛教中有两种禅修——一个是止，一个是观。"止"的禅修是心灵专注于一个对象上，不让心到处跑。"观"的禅修则是内观，将心向一切敞开。禅修能使有缘人可以根据自己的生活方式，随分随力地来提升自己的人生境界，减少贪欲、嗔恚、愚痴的束缚，逐渐让心灵净化并得解脱自在，使自己向善、向觉悟解脱的方向前进。

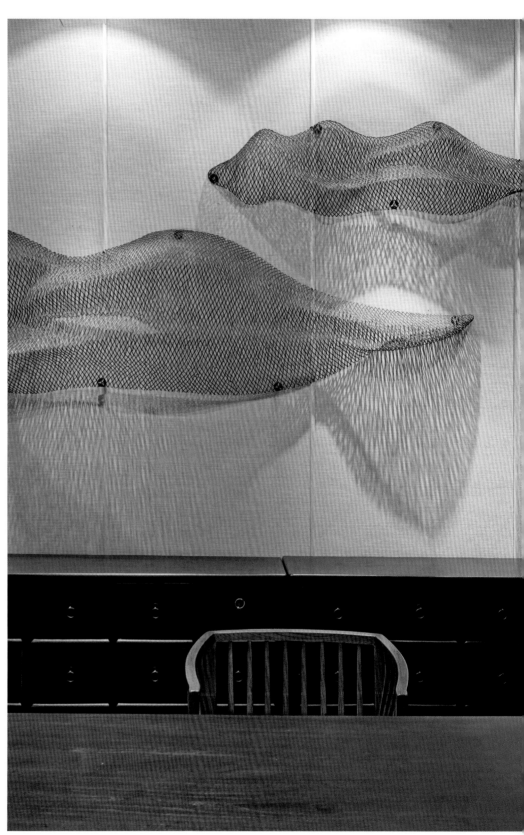

禅修会所软装概念定为"静"、"思"，用充满禅意的元素铺陈空间，使家具、配饰以及布艺的选择满足软装设计的需求。禅椅、云形装置、麻质布艺，美在还原本质的气韵，在纷繁的世界中寻找辨明自我的方式 —— "大道至简"。

The concept of furnishings in Zen meditation club is set as "peace" "thinking", to elaborate the space with the elements of Zen, to make the choice of furniture, accessories and fabrics meet the demand for furnishing design. The Zen's chairs, cloud-shape devices, linen fabrics, are so beautiful in restoring the original flavor, find the way of self-identification in the complex world.

设计师运用东方的极简手法来处理空间：包括柔和但富有层次的照明、最少种类的材质和米、灰、黑、白等接近无彩色的色彩搭配，以及减至不能再减的装饰。通过云、香、白沙、石等元素，营造出静谧得能让人忘却外界纷扰，放下一切回归本初的空间意境，达到让人自我修炼，关照内心的精神诉求。

The oriental minimalism is used in the whole design to deal with the space, and the lighting is soft but full of gradation, the least kind of textures, neutral-colored matching of beige, grey, black and white, and the decoration has been reduced to the minimum. Moreover, the elements like clouds, incense, white sand and stone are added to create a spatial prospect which is so peaceful for one to forget about the outside turmoil, put down everything and return to the prime.

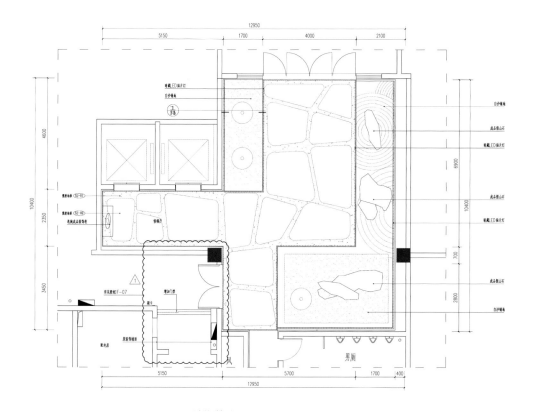

一层平面图
First Floor Plan

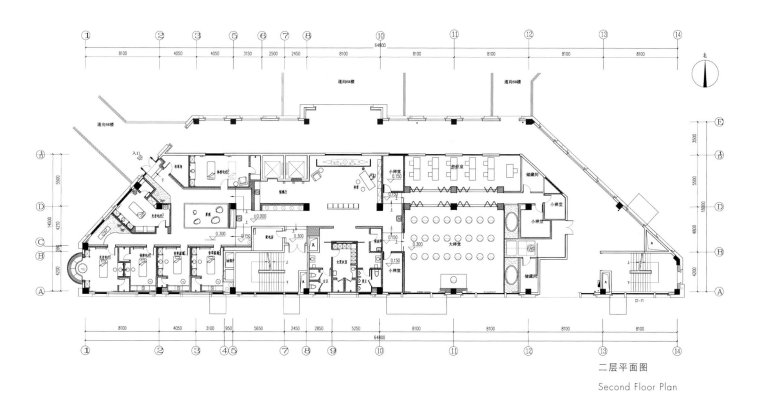

二层平面图
Second Floor Plan

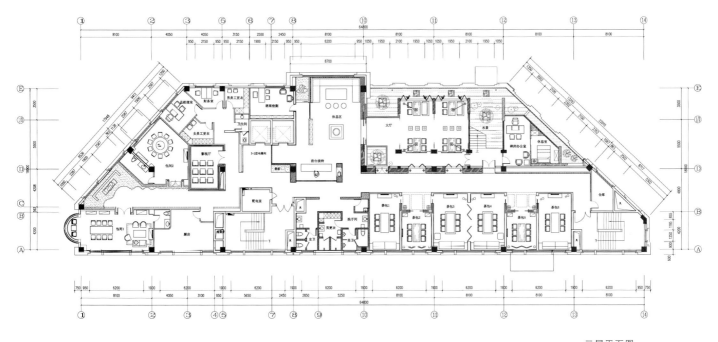

三层平面图

Third Floor Plan

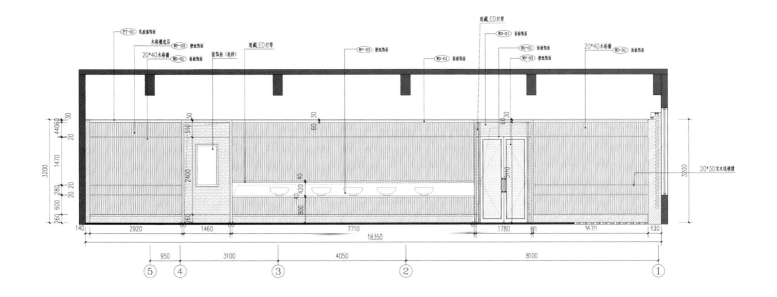

三层走廊立面图

Third Floor Hallway Elevation

国泰璞汇接待中心

The PUHUI
Reception Center

名称：

国泰璞汇接待中心

设计公司：

周易设计工作室

设计师：

周易

设计助理：

吴佳玲

摄影：

吕国企（和风摄影）

主要材料：

铁件、木格栅、水泥板、紫檀木、大理石、玻璃

Name:

The PUHUI Reception Center

Design Company:

JOY Interior Design Studio

Designers:

 Chou Yi

Design Assistant:

Wu Chia-ling

Photographer:

Hefeng Photography, Lv Guoqi

Major Materials:

Iron, Wooden grilles, Cement slabs, Rosewood, Marble, Glass

本案发想肇基于临时建筑如何低调融入周边地景，深度推演极简量体与环境的对应关系。建筑本体以方整的矩形打开横向面宽，设计上以简洁的水平、垂直线条结构，搭配不规则拼接的灰阶水泥板，展现建筑体外观的素朴与精致，更结合擅长的点状、带状情境光源，凸显绿地、水景，承托主建物的轻盈之美，隐喻内敛中蓄势待发的生命力。

The Project is based on how temporary building blends into the surrounding landscape, and deeply deduces the correspondence between minimum measurement and environment. The building has opened the horizontal width through the square-shaped rectangle, matched the grey cement with compact level and vertical line structure in design, to show the simplicity and delicacy of the architecture's appearance. Moreover, the design also highlights the green belt, waterscape to support the main building's beauty of lightness, and even metaphors the shovel-ready vitality with the adept punctiform and zonal ambilight.

几何延展的灰阶量体、左侧依适当比例浮凸利落的钢构、茶玻架构，宛若琥珀光盒般的意象，与水泥板的粗犷恰成生动对比。访客驻车后自树篱开口走上踏阶平缓渐升的抿石子步道，跟着迂回转折的步道灯光指引，可以欣赏沿途相随的无边界水池，安静地倒映漂浮水面的点点光影。主建筑物前端分别有两道呈90°角的屏风墙，运用细腻的墙体开窗方式，让视觉有条件穿透，赋予随机变化的框景效果。两墙中间围拱着前方圆形凿孔的天井廊遮，在这圆孔下方精心栽种绿竹，服膺取法自然的绿建筑概念，在视觉的导引上，则透过前后景的巧妙堆栈，兼具维护隐私与美化地景的实质意义。

The grey-scale building extended through geometry, the embossed and neat steel structure, tawny glass structure, prospect like the amber light box on the left side, have formed a vivid contrast with the roughness of cement slabs. After parking the car, the guest can step on the smoothly advancing pebble footpath from the quick-fence, and enjoy the borderless pond reflecting the little shadows on the water along the way with the guide of roundabout footpath's light. In the front of the main building, there are two 90-degree screen walls which can pierce the vision well and have a randomly changeable enframed effect with the smooth wall-windowing way. In the courtyard, there is a front gouge between both walls. The green bamboo has been elaborately planted on this gouge to bear the green architectural concept of taking advantage of the nature in mind, as skillfully stacking the foreground and background in the guide of vision can make the material meaning of both keeping privacy and beautifying the landscape.

推开特制竹编大门进入售楼处内部，亮黑色地坪延展的空间开阔而深邃，动线配置也如同简约外观的延伸。醒目的迎宾柜台是巨大的空间光点，由折纸概念而来的立体天棚与柜台基座，分别以木作搭配人造石建构，如同钻石立体切割的形体，在焦点灯光衬托下尤显力道遒劲，横斜其间的黑色树枝，别具设计者独有的风格印记味道。此外，柜台后靠的水泥板背墙穿插错落黑玻，除了让硕大墙体更有层次，也让墙后办公室内的工作人员可以透过玻璃窥视孔随时照看前台动静。

with wood and man-made stone, looking as if a diamond-cutting structure, which is especially powerful under the spotlight. The black branch stands in the structure is the designer's signature design. Moreover, the cement background wall behind the counter is interspersed with black glass, which adds up to hierarchical structure of the gigantic wall and allows the staffs behind the wall to see what is happening in the front desk through these holes.

When we push open the unique bamboo gate and enter into the interior, a widely open space with sparkling black flooring is in front of our eyes, and the movement arrangement is also as concise as its exterior. The eye-catching reception desk is a gigantic space light, and the three-dimensional ceiling and counter base stem from the concept of origami. They are built

柜台旁有方以大面格栅衬底的角落，用来展示兴建中的建筑模型，同样钻石立体切割的白色基座，透过上方自天花板深处悬垂而下的烟囱式聚光照明，凸显精妙的情境光源效果。

There is a corner underlain with large grilles opposite the counter, which is used to present the building models in construction. The diamond-cutting white base looks extremely exquisite under the chimney-style spotlight hanging from the ceiling above.

独立洽谈区的设计相当注重来客隐私，刻意降板的地面铺设长毛地毯，点缀其上的鼓凳带出微妙的东方人文，"ㄇ"型环绕的沙发同样低台处理，而嵌于沙发中央的装置艺术，利用相互衔接的亚克力棒，媒介上下光源的传导，表现烟雾般轻盈的光纤之美。设计者也特地在这里引入苏州庭园"有景则借、无景则避"的概念，配合横向大面玻璃窗，计划性地将窗外灰墙内的绿竹、光影意象吸纳入内，过滤多余街景杂质，仅保留室内人们立姿、坐姿时面向窗口最美的部份，也创造出一处气氛安全隐密却又无比舒适的洽谈环境。

The design of the independent discussion area focuses much on protecting the privacy of the clients. The deliberately lowered flooring is laid with long-haired rugs; the drum stools standing here present some subtle oriental humanity; the U-shaped surrounding sofa has also been lowered, and the installation art embedded among the sofa expresses the light beauty of fiber through those acrylic sticks. The designers specially introduce into this space the concept of Suzhou Gardens, by taking in the scene of green bamboos outside the window and the images of light and shadow, and filtering out useless street landscape. Only the most beautiful landscapes are left to be seen when people are sitting or standing in the area, which creates a private yet extremely comfortable discussion environment.

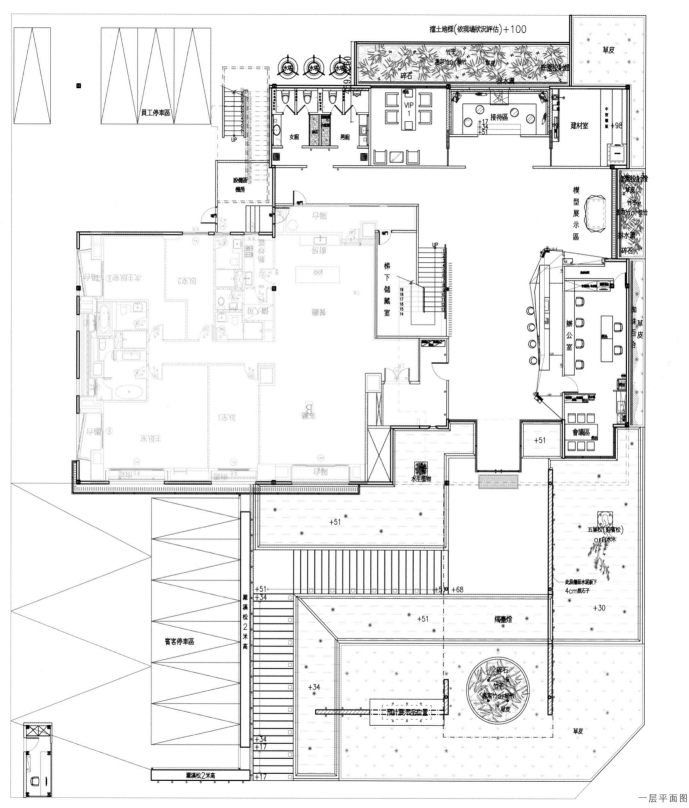

擋土地樑(依現場狀況評估)+100

草皮

員工停車區

女廁   男廁

VIP 1

接待區

建材室   +98

設備室機房

UP

模型展示區

梯下儲藏室

辦公室

會議區

+51

+51

水生植物

五葉松(迎賓松)
OF白木

+51

+5 +68

此段牆面水泥板下
4cm鵝石子

+30

+51   燭臺燈

賓客停車區

羅漢松2米高

+51
+34

+34

+34
+17

碎石
竹字
賓客竹or桂竹
草皮

+17

羅漢松2米高   +17

投影幕最佳位置

草皮

一层平面图
First Floor Plan

交椅
CHINESE ANCIENT FOLDING CHAIR

交椅，因其下身椅足呈交叉状，故名。交椅最早产生于唐代，是由胡床演变而来的。宋代时，交椅是一种比较高贵的椅子，只有仕宦大家或有名望的人才有资格置备交椅。它的特点是有轻巧的扶手，背板依照人的脊背作出曲线，座面是丝绳纺织的，颇为轻便舒适。其前后两腿交叉，交界点作轴，可以折合，上面安一固定圈儿，整个造型，从侧面看似由多个三角形组成，线条纤巧活泼，但不失其稳重。

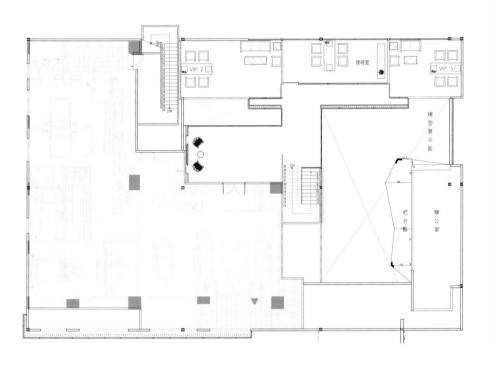

二层平面图
Second Floor Plan

太初面食

TAI-CHU Noodle Restaurant

名称:

太初面食

设计公司:

周易设计工作室

设计师:

周易

设计助理:

王思尹

摄影:

吕国企（和风摄影）

主要材料:

铁件、白水泥＋稻草、黑橡木、杉木、玻璃

Name:

TAI-CHU Noodle Restaurant

Design Company:

JOY Interior Design Studio

Designers:

Chou Yi

Design Assistant:

Wang Si-yin

Photographer:

Hefeng Photography, Lv Guoqi

Major Materials:

Iron, White cement + straw, Black oak, Cedarwood, Glass

这是周易设计团队的餐饮空间力作，为了传承祖业，并将深奥的面食文化发扬光大，身兼第二代经营者的店主与设计师周易再一次合作，希望赋予传统面食有如怀石料理般的精致与深度。

This is the masterpiece of dining space completed by Joy Chou's design team. In order to inherit the ancestral property and carry forward the deep wheaten food culture, the owner who is the second-generation proprietor has cooperated with the designer Joy Chou again, and expected to instill the delicacy and depth suchas Kaiseki cuisine into the traditional wheaten food.

在主述的环境氛围上，昭和时期日式民宅的怀旧与温暖以内敛、宏壮的建筑语汇加以呈现。线条极简的建筑外观大量采用特制铸铁与染黑的风化木，塑造出一种无畏变迁的强度，以及襁褓般的安定与温暖。而在近似灰黑色夜幕的结构背景里，设计师最擅长的塑景与情境光源营造技巧，透过点状、雾状光的描述，兀自挥发着水墨渲染的飘渺与力道。无论是店招上书法家李峰个性鲜明的道劲笔触，还是天然材质上的粗犷肌理，皆予人无穷深广的想象空间，并成功引导来客们的感官进入一个前所未有的人造臆境。

Under such circumstance, the nostalgic and warm feature of Japanese-style houses of Showa Period are presented here with reserved yet majestic structure, including the minimalist lines of the exterior, specifically made cast iron and weathered wood. All of these form the strength that stand against changes and the warmth as a cuddle. In the structural background of gray black nightfall, the landscaping and lighting design techniques that the designer is most adept has presented both lightness and strength as if ink wash rendering through dotted and fogged lights. Whether it is the calligraphy on the signage, or the rough texture of natural materials, they all spark people's endless imagination and succeeded in leading the clients' sensation.

原有的 4 米挑高骑楼增至 8 米，此举让寻常过道
有了殿堂的深邃与雄浑，光是在这处进入主空间
前的过渡地带，便汇聚了设计者的无数巧思。一
整段历经千百年风霜的粗壮原木，在这里反复搓
揉出各种姿态，首先是对剖后成为主水景要角，
虽是厚重木料却能轻盈浮托于水面之上，木头旁
成纵队的烛台灯与鸟笼，是设计者惯用的装饰元
素，呼应两列修长的竹管灯由上而下，以及视觉
可内外流动的格栅玻璃橱窗，低调解读多元文明
融合的人文之美。而格栅大窗上方的如折扇般的
曲面设计，加之向上投射光源的立体，着实为夜
晚增添无限情调。

The original four-meter high arcade has been increased to eight-meter high. This move allows the common pathway to have the depth and vigor of a palace. Light comes into this transitional area in front of main hall, which gathers numerous thoughts of the designer. The sturdy old logs are being made into various forms, such as the bisected log in the waterscape. Although the log is thick and heavy, it is able to float above water. The candle lights and birdcages lined beside the log are decorating elements commonly used by the designer, echoing two roles of slender bamboo lights in tube extending from the ceiling above, and the glass grid window that could flow from inside to the outside. All these elements interpret the humane beauty of a multicultural civilization. Moreover, together with three-dimensional lighting, the curve design above the grid windows also provide glamorous atmosphere to the evening scene.

走过横幅主水景进入餐厅之前，右侧另有一座精致水景，偏斜的日光映在铸铁墙面斑驳的锈雾上，上书"太初"的长筒状灯笼，悬挂在一方嶙峋的巴东石上，一旁玻璃橱窗内一块自大圆木支解的匾，彷佛是灯笼字迹的投影，更有店主无师自通的花艺画龙点睛，这样一个小品的角落，倏忽间就有了耐人寻味的怀石风情。

In front of the restaurant's entrance across the main waterscape, there is an exquisite waterscape on the right, with the deflective sunlight reflecting on the mottled rust of the iron wall and the long lantern which reads " 太 初 (Tai Chu)" hanging on the rugged Padang stone. In the vitrine by one side, there is a plaque divided from big round-log which seems to be the shadow of the lantern's handwriting, with the owner's self-taught floriculture being the highlights. This elegant corner has been filled with intriguing Kaiseki style in a flash.

入口处迎宾柜台立面外观与台面，甚至厨房送餐柜台的超长台面，同样引用大圆木的部份木材，整段木头串连整个空间，用法不同且没有丝毫浪费，这是设计师的巧思，更是对珍贵物料的一份由衷敬意。内部的用餐空间安静且素雅，视觉基本上是开扬的处理，交错融入东方的窗花画屏、日式的直列格栅，配合灰白泥墙、背景灯光等技法、素材运用，完成场域内虚实交错的界定，呈现出不同文化水乳相融的和谐，进入另一个层次。

The façade and table top of reception desk in the entrance, and even the overlong table top of meal-delivery counter in the kitchen, use the partial trunk of the big round-log to connect with the whole space, with different usages and without any waste. This is the designer's ingenuity, and even the sincere tribute to the precious materials. The interior dining space is quiet and elegant, with basically opening processing of vision. The oriental paper-cut painting screen, Japanese-style straight grid have been blended with the techniques and materials like ashen muddy wall and backlight matching up to complete the definition in the space, and make different cultures harmonious to get into another gradation.

隐约间可见餐厅后端另有一方不可思议的园林景致，地面铺上仿旧南方松地板，营造历尽风霜的朴实感，左侧石墙有水幕迤逦而下，然后汇集到"L"形的镜面水池。正前方刻意下压的窗景外，导入苏州庭园"有景则借、无景则避"的概念，一列修长绿竹迎风摇摆，水、石、竹、风、光五者缱绻纠葛，呈现极度迷人的深邃转折感，尤其在耳际萦绕淙淙的流水声里，更有极致脱俗的化外之美，深度撩拨人心。

Incredible garden view can be seen indistinctly at the back of the restaurant, as the ground has been paved with the southern pinewood to create the weather-beaten sense of plainness. The water on the stone wall on the left side is streaming down, and then gathers at the L-shaped mirror-surface pond. Suzhou Garden's concept of "to borrow the outdoor scenery if there is, to avoid it if there is none" has been introduced into the design of window scenery right ahead. A role of long and slim green bamboo is swinging in the wind, as water, stone, bamboo, wind, light are deeply attaching to each other to show the deep and charming sense of transition. In the bicker lingering in our ears, there is much more refined mystifying beauty that can deeply arouse one's heart.

❋ 黑色
BLACK

黑色在中原文化中被视为高贵的颜色之一，是政权、神权的象征。古代黑色为天玄，原来在中国文化里只有沉重的神秘之感，是一种庄重而严肃的色调，后来，它的象征意义受西方文化黑色贬义的影响而显得较为复杂。一方面它象征严肃、正义，如民间传说中的"黑脸"包公，传统京剧中的张飞、李逵等人的黑色脸谱；另一方面它又由于其本身的黑暗无光给人以阴险、毒辣和恐怖的感觉。

平面图
Plan

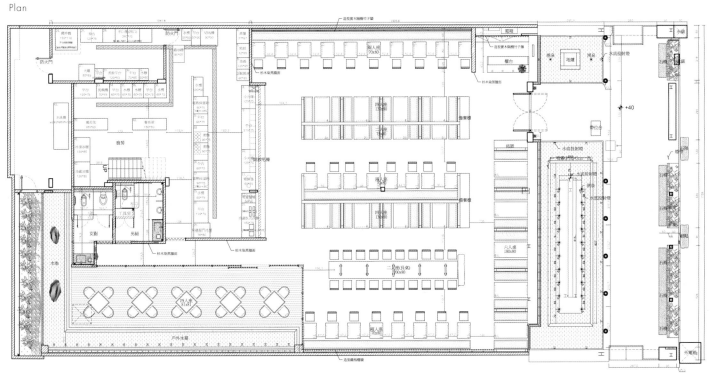

轻井泽公益店 & 三多店

Karuizawa Restaurant,
Gongyi Store & Sanduo Store

名称：

轻井泽公益店 & 三多店

设计公司：

周易设计工作室

设计师：

周易

设计助理：

陈威辰、陈昱玮

摄影：

吕国企、（和风摄影）

撰文：

林雅玲

主要材料：

公益店：铁件、铝格栅、文化石、铁刀木、南非花梨木、玻璃

三多店：铁件格栅、防清水模瓷砖、橡木染黑木皮、南非花梨木、南非
黑石皮、黑板石片

Name:

Karuizawa Restaurant & Sanduo Store

Design Company:

JOY Interior Design Studio

Designers:

Chou Yi

Design Assistant:

Chen Wei-chen, Chen Yu-wei

Photographer:

Hefeng Photography, Lv Guoqi

Article:

Lin Ya-ling

Major Materials:

Restaurant : Iron, aluminium grilles, Cultured stone, Venge, South
African rosewood, Glass

Sanduo Store : Iron grilles, Waterproof ceramic tiles, Black-dyeing
oak veneer, South African rosewood, South African black stone peel,
Black panel gallet

公益店
Gongyi Store

"轻井泽"公益店超过 7 米高的三面黑灰色建筑外观非常有特色，首先是大面铸铁精工打造的倒"L"形店招和雨遮，兼具等待区功能的木栈景观步道与建筑之间规划的镜面水景，点缀嶙峋的巴东石、烛台灯和蒸腾水雾，并贴心设置别致长凳以供来客小憩。最特别是在基座与马路的落差处，结合循环水幕与灯光，打造流动而梦幻的光瀑意象，夜幕时分，整座建物彷佛漂浮在光托之上。

Karuizawa Gongyi Restaurant's charcoal grey architectural appearance with the height of more than seven meters is very distinctive. First, there are the inverted L-shaped signage and rain awnings elaborately made with large cast iron, the jagged Padang stones are embellished by the mirror surface between wood stack footpath with the function of waiting area and architecture, along with candlestick lights and rising spray, and also unique benches have been set for the guests to take a rest. The most special thing is the drop between the pedestal and the street which creates the flowing and dreamlike image of lightfall as the whole building may seem to be floating on the light at night.

The black grilles with different sizes and long straight window have been set respectively on both sides of the building. Simplified grilling lines of the Chinese-style window paper-cut are blended into the architecture like an ancient warehouse or blockhouse. a Zen-Like environment, there are also the comfort of green bamboo and enthusiasm of Pinus morrisonicola flickering in the breeze.

建筑两侧立面分别装置大小交错衔接的黑色格栅以及纤长的直列开窗，在宛如古代仓库或碉堡的建筑中，融入中式窗花简化后的格栅线条，低调禅风里，同时摇曳着绿竹的写意以及五叶松的迎宾热情。

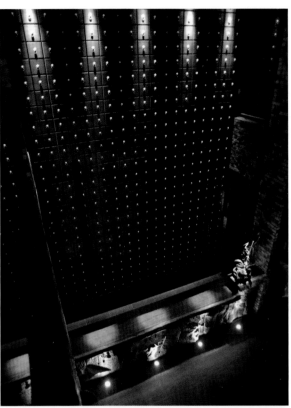

入口大门延续外廊的六角门拱设计，淡雅的苏州庭园意象隐约生成。内部分别为 4 米和 3 米的二层楼面，分别规划包厢与散座，总计 400 的座位，却能避免人声嘈杂、相互干扰，维持私密又有情调的餐叙氛围。负责全案主导的设计师周易曾说，"灯光是建筑的灵魂"一句话道破"光影决定情境"的真谛。

With the hexagonal gate arch spread through the entrance gate, the elegant scenery of Suzhou Garden has come into being indistinctly, as there are four-meter floor and three-meter floor which are respectively planned as box rooms and single seats. Although there are more than 400 seats, loud noises and disturbance can be avoided and the private and sentimental ambiance can be kept well. As the designer Chou Yi who has led the Project has expressed, "Lighting is the soul of architecture." This just really means that light and shadow can decide what the scenery may be.

内部空间设定延用大量实木格栅与情境光源，并融入处处可见的古玩书柜墙、大红色喜气洋洋的鸟笼灯饰，形成空间天井，并串连上下两个楼层的水景设计，架构三度空间，形成处处有景的绮丽空间。

The design of interior space uses a large quantity of wood grilles and ambilight, and visually added the antique bookcase wall that can be seen everywhere, the beaming red birdcage lighting into the space, to form a courtyard and connect to the waterscape designs on both up and down floors, and then build up the gorgeous three-dimensional space full of beautiful scenery.

🏵 中国灯笼
CHINESE LANTERN

中国灯笼是传统照明设备，起源于西汉时期。每年的农历春节，人们都挂起象征团圆意义的红灯笼，来营造一种吉利喜庆的氛围。中国式灯笼主要由纸或绢作为灯笼的外皮，骨架通常使用竹或木条制作，中间放上蜡烛，点燃蜡烛成为照明工具。灯笼综合了绘画艺术、剪纸、纸扎、刺缝等工艺，其中以宫灯和纱灯最为著名。现在华人社会中，灯笼已经变成了一种收藏与欣赏的传统工艺品，佳节庆典中都要使用灯笼，特别是春草、元宵节和中秋节。

多处墙面装置运用渐层玻璃与喷绘手法，透过晕染酝酿独特的"光迷幻"效果，舒缓而忠实地传达水墨书法与精致图章代言的人文精神。多层次、多角度的情境铺陈与灯光布局，透过多种自然材质与高难度工法，推出一幕跨越时间长廊的精彩演出，当然也让客人在就餐时有了更深层次的精神享受。

A number of wall installations use the gradient glass and airbrushed skill to express the cultural atmosphere through Chinese ink calligraphy and delicate seal, as the scenery in various gradations and lighting layout have formed an exquisite performance . This has certainly made the enjoyment with the five senses become much more profound.

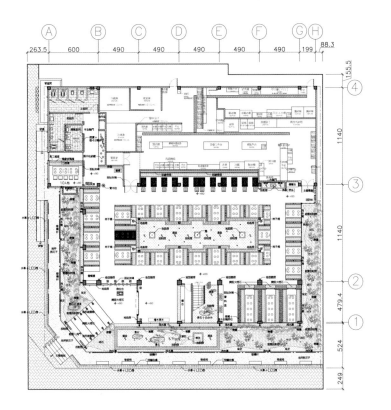

一层平面图

First Floor Plan

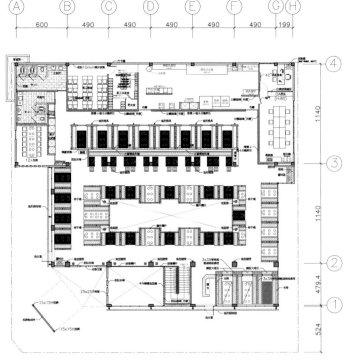

二层平面图

Second Floor Plan

三多店
Sanduo Store

远望朴雅宏伟的格栅外观，俨然是新东方美学最贴切的代言，建筑顶端以铸铁拉出轻薄屋檐，恰好挡住向上投射的灯光，整体采用沉稳的大地色彩，诠释安定的结构张力。

Overlooking the elegant and grand appearance of grilles, the store is just the most suitable symbol of New Oriental aesthetics, as the light eave made of cast iron on the top of the building just blocks the prejecting lights. The calm earthy tones have been used on the whole to explain the stable structural tension.

对开的桧木大门满是虫蛀形成独特的肌理，甚为珍稀，门旁一对业主珍藏的石狮分踞内外，取男主外、女主内的象征意义。视线沿着大块面连续的玻璃窗浏览，内部偏暗调的情境酝酿，与点状散布的灯光布局相辅相成。设计师通过局部斜角退缩的手法，于入口左侧特制一座能多方共赏的镜面灯光水景，搭配造雾器制造的氤氲效果，为池间一方拙雅凿石增添不少灵秀之气。

The folio cypress gate is full of moth-eaten texture, as two stone lions collected by the owner are sitting inside and outside of the gate, with significant symbol. Seeing through the large continuous glass window, the interior slight shaded scenery is fermenting, and supplementing with the punctiform dispersed lighting. Through the method of shrinking the local oblique angle, the designers has tailored a mirror-side lighting waterscape which can be enjoyed by various persons, with a fog-making device which is clouding on occasion to add the sense of delicate beauty for the quarry stone nearby the pond.

不同楼层的柜台设计，均采用同一设计手法，以岩石基座搭配实木台面，展现一种鬼斧神工的粗犷大器之美。一楼柜台后方悬挂一幅木雕的微笑佛陀脸谱，旁白"笑看人间的省悟"；二楼柜台旁修饰梯间的造型墙面，则以手工错落有致地绑上漂流木，呈现一种另类地格栅艺术。一楼后段的用餐区面向后院明亮的落地窗，窗外高至二楼的黑色木篱，圈出怡然自得的殊胜庭园，其间垫高的草皮小丘、斑驳的日影、精心摆放的灰白卵石，取法日式枯山水的自在禅意，早已渗入空气，与人们放松的心境合而为一。

The counter designs of different floors adopt the same design techniques, which apply rock base to match with the wooden table tops, showcasing the rustic beauty of artwork. The wooden-carved smiling Buddha mask is hanging behind the first floor counter, signifying the awakening of always looking at life with a smile. The wall beside the second floor counter is tied with many driftwoods by hands, presenting the unique art of grilles. The bright French windows in the dining area at the end of the first floor face the backyard, and there are black wooden fences with two-floor high that divides a beautiful space of comfort and carefree, where the greenery, lighting and cobbles are elaborately designed. The Zen spirit of Japanese-inspired garden designs here are already in the air, combining with people's relaxing state of minds.

三个楼层的营业空间都设置了舒适的包厢卡座，灰阶刻沟砖、大幅水墨背景、深色木格栅、镜面与椅背古老的车轹仔技法，在灯光的映射下，如老电影场景般展现。古朴静谧的情境，古老与现代交融的冲突演出，让人再三玩味。一楼角落转上二楼的梯间，阶梯踏面大理石与墙上凿面岩片彼此唱和，穿插渐层木纹板美化的梯线剖面，立面错置的凹格内点点烛光，在鲜明的光影阴暗交错间，放大写意的几何趣味。

The interior operation space is distributed on three floors. The comfortable balcony in division planning, are expressing the antique and fresh reminiscence like the old movies with the grey-scale carving groove bricks, or the large background of Chinese ink painting, deep-colored wooden grilles, mirror surface and old skills of axletree terminal at the back of the chairs

And the main scenery that is plain and peaceful sketches the contours of a kind of conflicted performance with the fusion of antiqueness and modernism to make one ponder over and over again. On the stairway from the second floor to the first floor, marble on the tread complements with the slice on the wall, as the ladder-tracked section beautified through the gradient grain board. A little candlelight in the uneven lattices dislocated on the façade have magnified the geometric interest of comfort in the lively dark shadow.

一层平面图
First Floor Plan

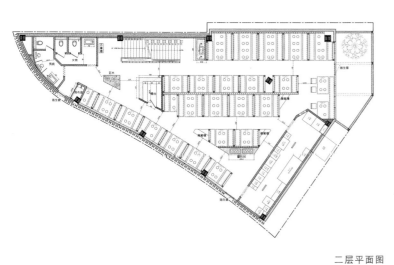

二层平面图
Second Floor Plan

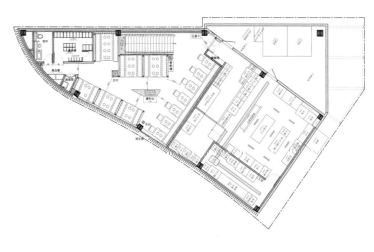

三层平面图
Third Floor Plan

这一锅朝富旗舰店 & 中山店

Top One Pot Chaofu Flagship Store & Zhongshan Store

名称：

这一锅朝富旗舰店 & 中山店

设计公司：

周易设计工作室

设计师：

周易

设计助理：

陈志强、徐嘉君

摄影：

吕国企（和风摄影）

撰文：

林雅玲

主要材料：

朝富旗舰店：铁件拼接板、仿古瓷砖、橡木染黑木皮、黑色烤漆玻璃

中山店：文化石、大理石、白水泥＋稻草、喷砂玻璃、黑橡木洗白、钢刷梧桐木

Name:

Top One Pot Chaofu Flagship Store & Zhongshan Store

Design Company:

JOY Interior Design Studio

Designers:

Chou Yi

Design Assistant:

Chen Chi-chang, Shiu Jia-jiun

Photographer:

Hefeng Photography, Lv Guoqi

Article:

Lin Ya-ling

Major Materials:

Chaofu Stone : Iron splice plate, archaized tiles, Black-dyeing oak veneer, Black paint glass

Zhongshan Store : Cultural stone, Marble, White cement + Straw, Sand-blasted glass, Black oak covered with white polyester paint, Steel brush, Platane wood

朝富旗舰店
Chaofu Store

建筑正面是采用金属结构、玻璃、麻绳为材质构建的细腻直列格栅，呼应飞檐底部向上投射的灯光，巧妙演绎古代帝王头冠前后摇曳的珠串印象。基座两侧以高低差打造沁凉的镜面水池，巨大朴拙的陶缸、水、灯光，成为灯光魔术师营造沉静优雅的禅意景观的最佳媒材。

The front of the building is using metal structure, glass and rope to jointly construct delicate lines of the grid structure, echoing the projection coming from the lighting on the bottom of cornice, which cleverly interpret the swaying back and forth bead lines of the emperor crown. On the two sides of pedestal, the high-low level difference is used to create a cooling mirror pool, huge and unadorned pottery vat, water and light have become the best media for the magician to create a tranquil and elegant Zen scene.

The dining areas on two floors inside are very spacious, with the sliding door set in a large quantity of box rooms to create the compartments, so that more flexible space distribution can be created if there are many guests. The dark tone is tranquil and solemn, so that the side view with the oriental charm can be presented strongly.

室内两层楼的用餐空间相当宽敞，多处包厢配置弹性拉门当隔间，来客多时就能做更灵活的空间分配。深沉的色调静谧肃穆，搭配更能衬托出东方韵味十足的端景。

天井两层挑高，于氤氲水景上凌空而起的转折光梯，每一阶都是深度加高度为 45 厘米的最佳尺寸，踏面内衬是近年时尚界推崇的客家花布，绚丽的色彩在灯光映衬下更显光华流转。

There is also the patio with two-floor height, the ladder of light up from the waterscape, and the golden ratio of the 45-cm depth and height on each step. Tread lining is the Hakka flower print that has become fashionable in recent years, as its gorgeous colors look much brighter in the light.

室内几乎每个角落的经营都是新东方美学的全新体验，如壮丽的云山图、白色浮雕窗花图腾、以诗经阳刻拓碑而成的局部立面造型、体态婀娜的描金仕女图、摆满拙雅铁壶的墙架、别致的齐白石水墨小品等。

The design in nearly all the interior corner is the new experience of New Oriental aesthetics, such as the drawing of Magnificent Yunshan Mountain, white totem of embossed paper-cut, local façade mould from The Book of Songs, drawing of noble woman with graceful posture, wall shelf displaying the refined iron kettles, Chi Baishi's unique Chinese ink wash painting, etc.

传统窗户
TRADITIONAL WINDOW

"窗"作为一种建筑上的构件，在实用性、装饰性以及象征性上都具有特殊的意义。传统窗户的样式有隔窗、短窗、支摘窗、横风窗、景窗和漏窗等。在初期，窗棂图案基本以规则的几何图案为主，如正方格。后来图案多样起来，还伴随着精美的雕刻，如人、神灵、花鸟等等。中国古代不同地区的建筑有着不同的风格，门窗的式样也情趣大异。比如侨乡福建和广东一带，人们喜好在建筑上大施彩绘，所以闽粤地区出产的门窗常见到有大漆绘就并加以纯金涂饰的。相反，色调以青灰为主的江南水乡，人们的审美观念中也以清新淡雅为美，因此门窗往往以木本色示人。

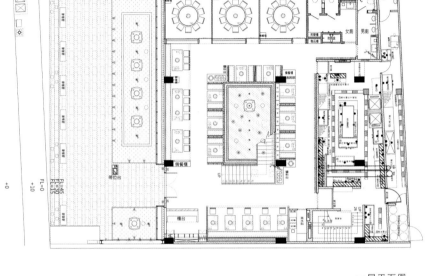

一层平面图
First Floor Plan

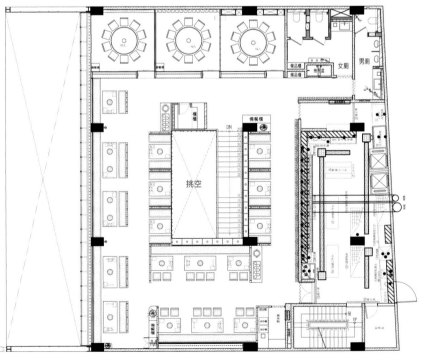

二层平面图
Second Floor Plan

中山店
Zhongshan Store

在骑楼走道的设计上，设计师大刀阔斧地将原地面的大理石及瓷砖全部挖掉，改以抿石子取代，并以弯曲流动的线条象征河流，结合华丽的波浪型天花浪板，使天地上下呼应，让骑楼走道展现出迷人的律动感。

The marble and tiles on the original ground have been all dug up drastically and replaced by the peddles, symbolizing the flowing river, and combined with gorgeous wavy ceiling to make the sky and earth complement with each other, and let the arcade's aisle show the charming sense of movement.

一棵长满树瘤的山茶花树干，除画龙点睛外，也是龙柱鼎立的概念延伸，其两侧做了两个堆满"柿子"的盆装造景，藉以象征"事事"如意、开店大吉。

A camellia burl is not only serving as the highlights, but also the extended concept of standing dragon pillars, while two basins of miniascape piled up with "Shizi (persimmon)" have been put on both sides to symbolize the wish of "Good luck in everything" "Get on well in business".

设计师周易采用雕花、窗花、花片、葫芦等具有中国东方特色的元素，结合书法笔墨之美，以及充满艺术风情的设计手法，打造一处尤如宫廷般的餐饮空间。

The designer. Chou Yi has used the elements as the symbols of Chinese oriental features like carving, paper-cut, flower petal, calabash, to create a dining space as the palace with the beauty of calligraphy and design skill full of artistic style.

采用典雅的窗框花片来分隔坐位，
既节省空间又可让每桌宾客享有私
密感。仅有的一间大包房，墙面是
借景中国苏州庭园柳树垂摆的景象，
结合梅花图腾的灯具来展现浓厚的
东方人文气息。

In addition, using the seats
separated by the elegant window
frame with flower petals can not
only be space-saving but also
allow the guests on each table
to enjoy the sense of privacy. In
the only one large compartment,
there is a prospect of the view
borrowing of dangling willows in
Chinese Suzhou Garden, and the
rich oriental humanities are shown
through the lighting with the totem
of plum blossom.

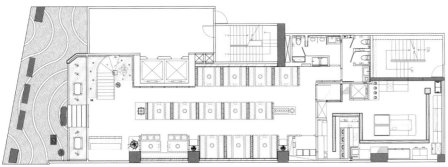

一层平面图

First Floor Plan

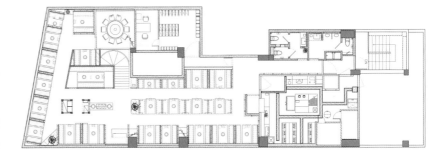

二层平面图

Second Floor Plan

深圳首誉销售中心

Shou Yu Shenzhen Sales Center

名称：

深圳首誉销售中心

设计公司：

戴勇室内设计师事务所

设计师：

戴勇

软装工程 \ 艺术品：

戴勇室内设计师事务所 & 深圳市卡萨艺术品有限公司

主要材料：

水泡石、比萨木纹云石、海贝花云石、地毯、钢化玻璃、
竹木饰面

Name:

Shou Yu Shenzhen Sales Center

Design Company:

Eric Tai Design Co., Ltd.

Designers:

Eric Tai

Decoration Engineering/Artwork:

Eric Tai Design Co., Ltd. & Shenzhen Katha Artwork Co., Ltd.

Major Materials:

Vesicular stone, wood-grain marble, seashell flower marble, carpet,
reinforced glass, bamboo ornament surface

本案的设计围绕着原项目名"芷峪澜湾"展开（后改名为"首誉"），希望能传达给客人自然宜居的项目形象。芷，本意为"香味令人止步的草"，隐喻花散发的香味；峪，即山谷；澜，水波，波纹；湾，水曲也，即河水弯曲处。设计传达了山谷的意念，利用曲线的设计语言，把山谷、花瓣及涓涓流水的联想视觉化，在现代语境里带出东方禅意。

Our design centers around the former name of the Project, ZHI YU LAN WAN (which was later renamed as SHOU YU), hoping to create an image of natural and liveable development. ZHI, originally referring to "grass making people stop their steps with its fragrance," a metaphor meaning fragrance of flowers; YU, a valley; LAN, waves, and ripples; and, WAN, winding water flow, or the winding point of a river. The design conveys the image of a valley, and presents the vision of valley, flower petals, and winding rivers by way of designing language using curves, thus bringing in the oriental image in modern language.

以花瓣为主题的发光墙体改善了原有封闭建筑空间的不舒适感，并传达出鲜花散发出清香的概念。天花的竹木曲线造型，形成犹如溶洞中即将滴落的石钟乳，在自然重力的作用下垂落到地面的视觉效果，巧妙而又自然地隐藏起原建筑结构中六根大体量的结构柱。从天花上垂下来的琉璃吊灯，像滴落的水珠被定格在半空中，晶莹闪烁。

Lighting wall body with flower petal theme improves the sense of unease in the originally confined building space, and exhibits the vision of flowers giving off fragrances. The patterns of bamboo curves on ceiling presents the vision of stalactites about to drop off in the cave, thus hiding naturally the six mega size structural pillars in the original build. The glass pendant lights from the ceiling look as if the drops of water setting in the sky, crystalized and shiny.

一叶一菩提，一花一世界

TO UNDERSTAND BODHI IN A LEAF, AND THE WORLD IN A FLOWER

源于佛教经典华严经，这个经典源于一个故事：佛在灵山，众人问法。佛不说话，只拿起一朵花，示之。众弟子不解，唯迦叶尊者破颜微笑。只有他悟出道来了。宇宙间的奥秘，不过在一朵寻常的花中。这昭示着佛是何以知道有微观世界，何以知道有宇宙的。研究一朵花，可以悟出点东西，此所谓道，然后"道生一，一生二，二生三，三生万物"，我们了解的便会趋于无穷。

平面图
Plan

自然是构成意象的根本，设计师摒弃了销售中心一贯体现奢华的设计手法，而是运用主题切入的手法，让每个造型元素都为设计概念服务，营造出清新脱俗、清澈自然的空间效果。

The nature is the basis for vision and imagination. The designer abandons the design techniques commonly used on sales center to present luxury, but instead, he applies the theme into the design, making each element to work for the design concept and ultimately presenting the refreshing and natural effect.

写意东方

FREEHAND EAST

名称：

写意东方

设计公司：

唐玛国际设计

设计师：

施旭东

摄影：

吴远长

主要材料：

镜面拉丝黑钛金、黑金龙大理石、皮革编织装饰、蘑菇面
芝麻灰花岗岩、黑镜、银色仿古镂空画、壁纸、擦色欧饰
墙板

Name:

FREEHAND EAST

Design Company:

TOMA

Designer:

Shi Xudong

Photographer:

Wu Yuanchang

Major Materials:

Wiredrawing black titanium of mirror surface, Black Gold Dragon
marble, Leather-woven adornment, Sesame gray granite with
mushroom surface, Black mirror, silver archaized pierced drawing,
Wallpaper, Brush-off European-style wallboard

黑金龙大理石地面，黑灰色如流水般的纹理从电梯间开始延伸，光洁的大理石质地与柔顺的流水纹路碰撞出双重的美感。中国"道德经"的雕刻字块被设计师当做设计元素，排列在墙面上，四大发明的铅字印刷被以现代材质完美演绎。造型夸张，强调尺度感的中式椅子，用亚克力来表现。现代材质与传统元素相得益彰，既有中国传统神韵，又有强烈的现代感。

On the ground with the Black Gold Dragon marble, the black gray grain like the flowing water has extended from the elevator room. The marble's glabrous texture and smooth grain like flowing water have colloded into the dual aesthetic feeling. The type printing in China's Four Great Inventions has been shown through the modern textures, as the carving characters of " 道 德 经 (Tao Te Ching)" flatly sorted on the wall by the designer has an overstated shape. The Chinese-style chairs, which stress the sense of dimensions and are expressed with the acrylic. Modern materials and traditional elements are complementing each other, providing traditional Chinese artistic sense to the space, as well as strong modern aesthetics.

传统紫砂工艺"提梁壶"被大胆"破坏"，以当下的设计形式演变成既具功能性的接待前台，又有装饰性的纯白色当代装置艺术品，静立于大厅中。摇曳的烛光、朴拙的石材栓马桩、金属装饰品，以及青花图案，设计师利用色彩、材质、光影以及造型的穿插、对比、协调所产生的张力，来引起来宾的共鸣，传达出空间要述说的东方内在精神。

Chinese traditional art of polished lacquer painting depictsthe theme of "lotus", mottled with strong red lacquer. The loop-handled teapot of traditional red porcelain has been "damaged" daringly. There are also flickering candlesticks, plain stone pitching post, metal decoration and the patterns of blue-and-white porcelains. The designer has given rise to the readers' emotional resonance with colors, textures, shadow and strain created from the formative interspersion, contrast and harmony, to express the oriental inherent spirit.

❋ 提梁壶
SWING-HANDLED TEAPOT

提梁壶是指以提梁为把的紫砂壶，壶式之一。始于北宋，流行于明清。小口、细流，鼓腹，平底，有盖，为了提拿方便，在肩部两端连以半月形提梁，宋代耀州窑创制，明清均有烧造，品种有青釉、青花、粉彩、紫砂等。提梁壶可分为两种形式，一种是硬提梁，另一种是软提梁。民间流传提梁壶是由苏东坡所创造，并把这种式样的茶壶叫做"东坡提梁壶"，或简称"提苏"。

一颗颗珍珠透过玻璃和光影的洗礼，有序地拉出优美的弧线。包间主题背景墙是以中国传统磨漆画技法描绘的"莲"，浓烈的红色大漆，饱满深沉而斑驳。

The pearls pull out pretty arc one by one through the glass and shadow. The current design forms have been changed into the functional reception desk and the contemporary pure white decorative artworks in the lobby.

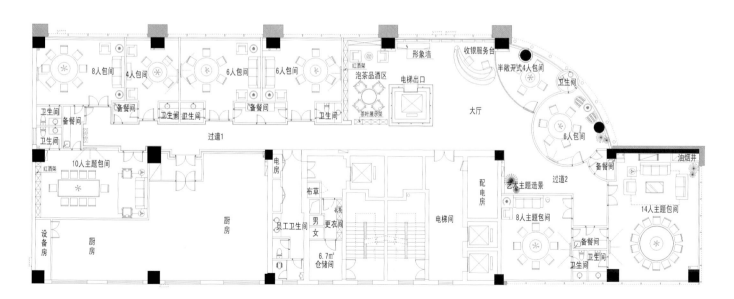

平面图
Plan

工艺美术
Arts and Crafts

工艺美术

在古代，工艺美术就是含有艺术价值的手工制品。它的生产形态等同于手工业，它的文化形态则属于造型艺术。"工艺美术"是个后起的词语，在中国的历史尚不足百年。该词语出现得虽然晚，但它涵盖的若干门类却是最早的艺术创造。值得相信的是，在中国的原始社会，工艺美术比美术更成熟、更辉煌，而且原始的绘画和雕塑也常常依附在工艺美术作品之上。

玉器在中国美术史上占有重要的地位，因为它是由石器演变而来的，使人想到了遥远的石器时代。《说文解字》上这样定义玉的概念："玉，石之美者也。"中国人对玉可以说是极其偏爱，从兵器变成礼器、变为丧葬用物、变为装饰文玩，一直到现在玉仍是大多数中国人的爱好之物。而古玉的形制名称亦非常复杂，非常值得推敲研究。以圆环的玉璧为例子，《尔雅·释器》上说"肉倍好，谓之璧。好倍肉，谓之瑗。好肉若一，谓之环。"即是洞眼最大的圆形玉叫做瑗，洞眼适中的叫环，洞眼较小的就是璧了。

汉代南越王墓出土的玉璧

在清代以后，和阗的软玉和缅甸北部的硬玉（翡翠）大量涌入中原，玉的纹饰雕琢更加精美，所以在汉代以前的玉器讲究的是时代和形制，明清以后，讲究的就是纹饰和雕琢了。

鎏金舞马衔杯银壶

金银器物在唐朝得到了极大的发展，唐代是金银器佳作纷呈的时代，作品格调高雅，花纹生动，洋溢着一种后世难以企及的美感。由于丝绸之路的持续畅通，6世纪初期的金银器物受到了拜占庭、萨珊王朝栗特的影响，直到7世纪晚期，唐朝的金属制造方才脱离外来的器形和纹饰的影响，唐朝后半段达到了金银器制造的顶峰。从出土于西安南郊何家村窖藏的鎏金舞马衔杯银壶中，我们就可以一窥盛唐的气象。

陶瓷是中国历史发展过程中不可或缺的重要工艺制品，更是中国文化的一项重要特色，在英文中瓷器和中国是同音同义的词，在外国人眼中，瓷器就等于中国，可见瓷器艺术在中国的特殊地位。

古代中国人先发展了陶器，又由陶器发展出了瓷器。所以中国瓷的历史是从中国陶开始的。仰韶文化、马家窑文化是彩陶的代名词，而山东的龙山文化则是黑陶的代名词。在魏晋南北朝时期，出现了原始的瓷器。

而隋唐则是中国瓷器的成长期，形成了"南青北白"的局面。邢窑白瓷大为流行，与越窑青瓷一起作为当时并重的两种瓷器，白居易曾歌咏白瓷"白瓷甚洁，红炉炭方炽"。唐代盛行饮茶风尚，茶器成为必不可缺之物，有力刺激了唐代瓷器的发展。白瓷虽也作茶具，在唐代却非主流，远不及时人对越窑青瓷的喜爱。这是因为唐代罕见白茶，茶以绿色为主，绿色之茶与青瓷交相辉映，有"益茶"之效。

唐代的邢窑和越窑都很有名，但是盛唐时期盛极一时的唐三彩器皿却是中国陶瓷史上的一朵奇葩。它是一种以低温烧成的彩色釉陶器，主要为冥器。它因釉色多为黄绿白，所以称之为三彩，其实其远不止此三种颜色。其釉色具有变化莫测、淋漓流淌的自然而奇妙的效果，红极一时。

唐三彩

元青花鬼谷子下山罐

宋瓷是中国瓷器造诣的高峰，主要有汝窑、哥窑、官窑、钧窑、定窑五大窑以及龙泉窑。而拥有最高等级瓷土的景德镇瓷窑也是在宋代早期建立发展而来的。

明代则是以青花瓷为特色，这种釉下彩的青花瓷最为辉煌的时期，当属明代宣德和成化两朝。前者以它的形制和釉色闻名于世，后者则因为它的彩绘不但画笔灵动而且斗彩明艳，而成为一代佳作。

清代的瓷器制作在技术方面达到了登峰造极的境界，不但可以仿古，而且由于西方珐琅技法的传入，又在瓷器上大放异彩，清瓷最盛为康熙、雍正、乾隆三朝，最高成就则是珐琅彩瓷。珐琅彩瓷的绘画是其精华所在，大多出自宫廷御用纸绢画的画稿。

掐丝珐琅兽面纹提梁壶

中国的工艺美术，历史悠久，品种繁多，技艺精湛。它蕴含着中国人民的智慧，融汇了中华民族特有的民族气质和文化素养，以其生动的神韵蜚声国内外，是世界文明中一颗闪光的明珠。

工艺美术作品的内涵是非常丰富的，形式也是多样的。在室内设计当中，艺术品的设置主要由室内空间的用途和性质决定。使艺术品和室内环境融合衬托，交相辉映，才能使装饰因素与结构因素、构造因素紧密地结合在一起，形成一个比较完满的整体。

康桥半岛私宅项目

Cambridge Peninsula Residence

名称：

康桥半岛私宅项目

设计公司：

西玛设计工程（香港）有限公司

设计师：

郦波

摄影：

谢国建

主要材料：

玫瑰钛金、黑色钛金、白色钢琴漆、实木花格、柚木、黑檀木、
玻璃、黑木纹大理石、墙纸、定制地毯

Name:

Cambridge Peninsula Residence

Design Company:

CIMAX Design Engineering (Hong Kong) Ltd.

Designers:

Li Bo

Photographer:

Xie Guojian

Major Materials:

Rose titanium, Black titanium, White piano lacquer, Wood lattice,
Teak, Ebony, Glass, Black wooden marble, Dark wooden wallpaper,
Custom carpets

本案业主是位优雅而具中国书卷气质的女士，既喜欢美式的舒适与温馨，同时又钟情于收集中国情趣玉石、奇石、器物。对于本案的生活空间，业主希望能够既兼容美式的舒适性，又能兼具中式的时尚文化韵味。

The owner is an elegant, well-educated Chinese lady. As well as having an interest in the American life-style, she is an avid collector of interesting Chinese jade, exquisite rocks and artifacts. For her living space, the owner wanted to be able to have the comforts and warmth of the American life-style while still enjoying the charms of traditional Chinese fashion and culture.

设计师用一些新的形式和载体来恰如其分地表达中式韵味和文化，尝试去探求一些语言、形式和载体上的变化，从一些特殊的视角还原真实的中国趣味，将中国味道表现出时尚与文化共融的新式优雅，以极其浓厚的色彩诠释空间和物型，让空间生动、丰富。灯光、阳光以及园林设计外景内引，在不同的时段、不同的光线和不同的色彩会创造出不同的空间氛围，使得整个空间的色彩更加生动而富有变化，完美演绎了"家"的情趣。

In order to express, explore and highlight the language, form and manner of Chinese cultural aesthetics in a gracious way. We've used strong colors to interpret the living spaces, making them much more vivid and rich. The lighting, use of natural sunlight and the interior landscape are all combined to create a special atmosphere at different times of day, with the changing of colors and lighting we can achieve a much more colorful, vivid, rich and convivial expression of " home".

对于公共空间，设计师在餐厅与走道之间以一组透明亚克力材质的弧形展示柜做隔断，灵动地进行了光线的沟通，又因为形式和载体的改变，使博古架摆脱原有的一贯形象，显得更加时尚，而透明的材质更好地将厚重的藏品全面地展示出来，熨帖巧妙。

In the public living areas, the designers set a group of transparent acrylic display shelves between the walkways and the dining area, the lighting has been cleverly manipulated in such a way that it moves by changing its form and carrier, highlighting all the pieces on display and in a stylish way making them stand out even more.

娱乐室的中式圆形花格窗，其倒影巧妙地借景了园林中的水镜反射。经过设计师精心的嫁接，水墨背景图、祥云凳、格栅门等东方元素，配衬中西结合的家具，使东方意境恰如其分地融入了设计师要讲述的故事里。

Traditional Chinese latticed round windows in the entertainment room form a tasteful reflection of the garden landscape in the pool of water. The entire entertainment area is a fusion of both Western and Chinese furniture (ink painted slice door, stool with cloud patterns, grilled gate etc.) all are appropriately arranged to infuse an overall oriental theme and mood.

根据业主对空间私密性的要求和生活喜好，设计师在书房区域独辟出清宁、时尚的中式茶室，并在室外园林搭配一棵银杏树，与院墙的一排青竹相互辉映，让茶室更清幽。红色丝绒陪衬钛金搁架和白色壁笼，以及白色中式屏风，营造出书房的重器收藏品的雅致。

Based on the owner's preferences, the designers made a clean, stylish Chinese teahouse in the study room with an outdoor garden lined with ginkgo and bamboo to extend the tearoom's tranquility. Red velvets set off the titanium foil frame, white walls and white Chinese screens, creating an elegance that highlights the study room's valuable collections.

依循业主对于私密空间和公共空间的要求以及生活习惯，设计师在本案的功能动线和使用细节的设计上，合理地兼容了美式重视生活自然舒适性的特点，满足了业主对空间要个性、实用、舒适的要求。

According to the owner's requirements on private and public space and living habits, the design of the function movements and use details are reasonably merge with the natural and comfortable American style interior, which perfectly meets the owner's requirements on personality, function and comfort.

主卧区域的梅花喜鹊报春床屏与荷花屏画，陪衬真丝织物床品和定制的泼墨地毯，彰显出业主的清雅个性，与私密休息区域的玉兰织锦屏风、钛金搁架和内嵌式壁炉相得益彰。

In the master bedroom, we focused on highlighting the owner's elegance by using a tasteful plum blossom and magpie screen above the bed, a painting of lotus flowers on the wall, silk fabric bedding and a custom inked carpet. In the private lounge area, a magnolia tapestry screen and titanium shelves are perfectly complemented by the fireplace.

✿ 屏画

PAINTING ON PARTITION SCREEN

建筑中室内障子、屏风、屏障上绘画的总称。以木结构建筑为主的建筑室内的隔断、间隔的墙壁大都有各种形式的活动壁、活动拉门、屏风、影壁等，在这上面大都绘有各种题材的屏画。屏画最早可追溯到古坟壁画，在日本艺术界得到极大的发展。中国绘画中还特有"重屏"这个题材，尝试制造视觉幻象，采用相同的透视，诱使观者将屏幕的画当作现实。

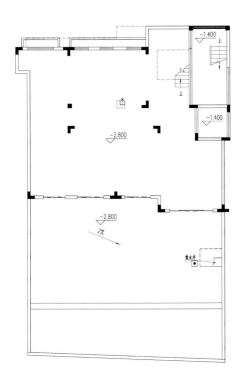

地下层原始平面图
Original Basement Plan

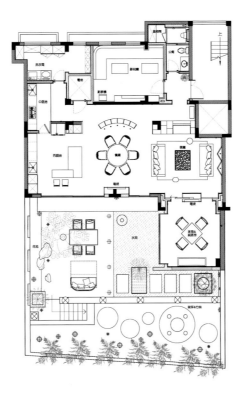

地下层新平面图
New Basement Plan

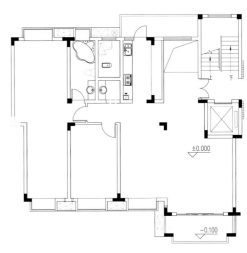

一层原始平面图
Original First Floor Plan

一层新平面图
New First Floor Plan

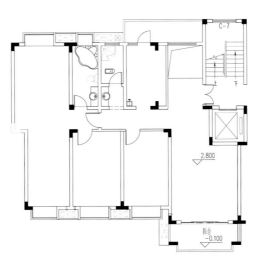

二层原始平面图
Original Second Floor Plan

二层新平面图
New Second Floor Plan

兰亭壹号

No.1 Lanting Club

名称：

兰亭壹号

设计公司：

河南鼎合建筑装饰设计工程有限公司

设计师：

孙华锋、孔仲迅

摄影：

孙华锋

主要材料：

黑白玉石材、法国木纹石，红橡面板、草编壁纸、实木花
格等

Name:

No.1 Lanting Club

Design Company:

Dinghe Architectural Decoration Design Project Co. LTD

Designers:

Sun Huafeng, Kong Zhongxun

Photographer:

Sun Huafeng

Major Materials:

Black & white jade stone, serpeggiante, red oak panels, straw-
woven wallpaper, solid wood lattice, etc.

兰亭壹号位于山西太原滨河东路，是以尊崇传统文化及健康养生经营理念为主导的高端餐饮会所。水元素的加入活跃了空间，潺潺细流的水幕墙背后是雕刻着王羲之《兰亭集序》的石板，静逸中透着灵动。走廊尽头万佛墙与现代工艺制作的金属荷花的搭配，材质冲突中意境却和谐，呼应了传统结合现代的设计手法。

Located on Binhe East Road, Taiyuan, Shanxi, No.1 Lanting Club is a high-end catering club in pursuit of traditional culture and health preservation as its core philosophy. By adding the element of water, the space looks more vivid. Behind the water-curtain wall is a stone tablet engraved with the Preface of Orchid Pavilion written by Wang Xizhi, achieving a perfect infusion of tranquil and vibrant elements. With the combination of Wall of Ten Thousand Buddha at the end of the corridor and the metallic lotuses made with modern technologies, the different textures give us a sense of harmony, echoing the combination of traditional and modern design styles.

室内空间融中国传统文化的高
贵典雅与现代时尚创意于一体。
在运用现代设计语汇进行室内
空间规划的同时，设计师运用中
国古典窗格、传统工艺的漆柜等
元素提炼出具现代审美情趣的
形式，使其成为空间界面的肌
理，透过光线的映衬烘托出静
谧、深沉、高雅的空间氛围。

Its interior space presents a perfect
infusion of the elegance of Chinese
traditional culture and modern fashion
and creativity. While planning the interior
space with modern design vocabularies,
the designers utilize such elements as
China's classic traceries and lacquered
cabinets of traditional technology to
refine and obtain a modern aesthetic
form. These elements become the
textures of space interface, and form
a space atmosphere of tranquility,
deepness and elegance in the light.

会所的陈设设计采用中西合璧的方式，力求通过将西式家具的舒适度与中式家具的传统韵味相结合的方式营造私密尊贵且具人文气息的空间气质。墙面悬挂的字画均为大师真迹且经过设计师对比例尺度的严格把控，使其和谐融入空间之中，是整个会所的点睛之笔。精心定制的灯具与空间形成完美的对话，传统绘画复制的漆屏散发着细腻悠远的古韵遗风。

By adopting Chinese and western design styles, namely, combining the coziness of western-style furniture and traditional charm of Chinese style furniture, the club's space is characterized by privacy, distinguished and humanistic culture. It's the designers who strictly controlled the ratio scale of the calligraphy and paintings that are hung on the wall, known as the masterpieces of the maestros, so that the calligraphy and paintings can be integrated into the space harmoniously, becoming a core part of the whole club. The custom-made lamps form a perfect dialog with the space, while the lacquered screen decorated with traditional duplicative paintings present an exquisite and antique charm.

❀ 虢国夫人游春图
PORTRAIT OF LADY OF GUO
ON A SPRING OUTING

《虢国夫人游春图》，为张萱名作，现存为宋徽宗赵佶摹本，原作已失，摹本犹存盛唐风貌。为流传有绪的唐宋名迹中稀有瑰宝之一。作品描绘了杨贵妃姐妹三月三日游春的场景。画作线条简劲流动，敷色艳丽鲜明，人物体态丰满华贵，具有鲜明的时代特征。

整个空间沉而不闷、透而不散，于无形间流露出浓浓的古典人文气息，表达出设计者对传统文化的敬意与向往。置身兰亭壹号，享受美食与艺术的美妙体验。

The whole space is deep but not dreary, transparent but not scattered, revealing classic humanistic culture, and delivering our respect and aspiration for traditional culture. Welcome to No.1 Lanting Club to enjoy a wonderful experience of cuisine and art...

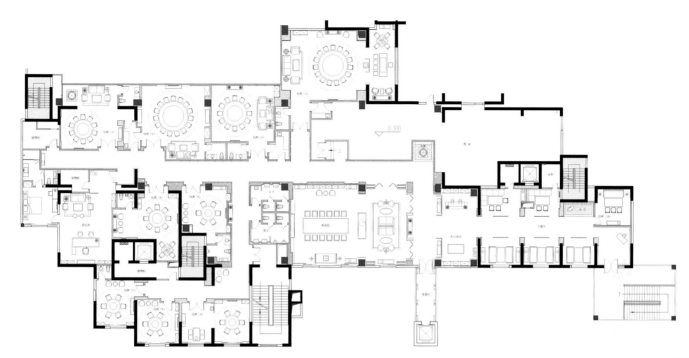

平面图
Plan

墨 韵

Charm of Ink

名称:

墨韵

设计公司:

唐玛国际设计

设计师:

施旭东

摄影:

周跃东

撰文:

朱含薇

主要材料:

木纹大理石、磁木系列、编织文硬包

Name:

Charm of Ink

Design Company:

TOMA

Designer:

Shi Xudong

Photographer:

Zhou Yuedong

Article:

Zhu Hanwei

Major Materials:

Marble with wood grain, Magnetic wood, Hard clad with woven patterns

沿着楼梯婉转而上，错落有致且形态各异的金属吊顶、木色伞骨架和青砖石瓦般的四壁仿佛瞬间将人们带入了一幅东方命题的画卷。在这个充满想象的空间里，精神和意境、品质与灵魂、当代艺术和传统文化浪漫邂逅，一如设计师一贯的风格：强调新东方精神，强调艺术与空间的碰撞。通过残破的、被剥离的传统符号的抽象运用，用现代手法和工艺营造东方文化的艺术空间。

When going up tactfully through the stairways, the well-arranged metal ceilings in different shapes, wood-colored skeleton of umbrella and walls of blue bricks and tilestone seem to bring one into an oriental painting. In this space full of imagination, spirit and prospect, quality and soul, modern art and traditional culture romantically meet, as the designer's consistent style-emphasizing the spirit of New Orientialism, collision of art and space. Through the abstract use of wrecked, stripped traditional signs, the artistic space of oriental culture is created with the modern techniques and crafts.

✿ 油纸伞
CHINESE TRADITIONAL UMBRELLA (OIL-PAPER UMBRELLA)

油纸伞是中国传统伞，作为起源于中国的一种纸制或布制伞，亦传至亚洲各地如日本、朝鲜、越南、泰国等地，并发展出具有当地特色的油纸伞。油纸伞除了是挡阳遮雨的日常用品外，也具有特殊的象征含义。中国传统婚礼上，新娘出嫁下轿时，媒婆会用红色油纸伞遮着新娘以作避邪。宗教庆典中，也常看到将油纸伞作为遮蔽物撑在神轿上，取其圆满的意思，作为人们遮日避雨、驱恶避邪的象征。

大堂中四个姿态不一的红色人物雕塑，设计师参考了唐朝《舞乐屏风》——以舞伎、乐伎为制作题材，锦袖红裳，人物飘逸优美。"前音渺渺，笙箫笛筝，琵琶拍板，筚篥鼓叶"，这组乐伎雕塑堪称中国古代贵族生活的传神写照。荷花元素在立体壁画和屏风中被大面积地使用，这出"淤泥而不染"的清廉之花深得古代文人雅士的喜爱，但从设计师手中表达出来，却多了一丝婀娜多姿的自然之趣和大气深邃的东方意境。

In the lobby, four red figurines with different poses are just like a screen, as the designer has referred to Tang's Wuyue Pingfeng-with the theme of dancer and musician, graceful and pretty women in red brocade dresses. "Visionary sound, from the flutes, from the lutes, and also from the Tartar pipes". The set of Yongling's musician sculpture are just the portrait of Chinese ancient noble's life. The lotus is massively used in the solid wall painting and screen. This pure and clean flower has gained the love from the ancient refined scholars, but it has also been given a coquettish pleasure of nature and gorgeous oriental prospect by the designer's expression.

风格是艺术的生命力，传统东方文化、艺术空间的呈现绝不是符号的简单罗列式复制，而是追求在文化、艺术的前提下，通过当代设计形式、语言表达当代人的审美气质。在彰显典雅与品位的同时，更多的是对"人"的尊重，而不是简单的形式的表现。设计师利用色彩、材质、光影以及造型的穿插、对比、协调所产生的张力，传达出空间要述说的东方内在精神，引起人们的情感共鸣。

Style is vital to art. The presentation of traditional Oriental and art space is absolutely not the simple enumerating copy but the expression of contemporary people's aesthetic through the current design language of culture, art, black and white. It shows the respect for "human" with the elegance and taste instead of simple expression of form. The designer has expressed the oriental inherent spirit with colors, textures, shadow and styling created from the formative interspersion, contrast and harmony, to give rise to people's emotional resonance.

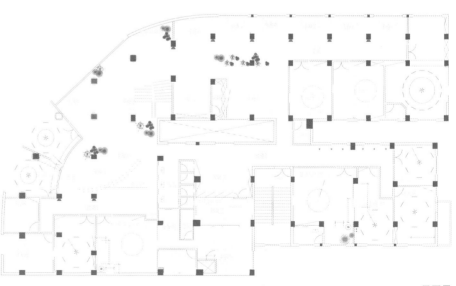

平面图
Plan

豪笙印溢

Hao Sheng Yin Yi

名称：

豪笙印溢

设计公司：

美国 IARI 刘卫军设计事务所

设计师：

刘卫军

Name:

Hao Sheng Yin Yi

Design Company:

Liu & Associates (IARI) Interior Design LTD.

Designer:

Danfu Liu

本案以简洁、明晰的线条和优雅、得体的装饰，展现出空间中华美富丽的气氛，表达了一种随意、舒适的风格，将家变成释放压力、缓解疲劳的地方，给人以典雅宁静又不失庄重的感官享受。简洁的天花设计与自然木纹形成的几何图案地板相呼应，把握了美式风格的简洁、对称、幽雅的精髓，空间中白色大理石立柱的运用则表达了一种更加理性、平衡、追求自由、崇尚创新的精神。

The Project shows the gorgeous ambiance in the space with the concise, clear lines and elegant, graceful decoration. It expresses a casual, comfortable style, which has changed the home into a place for releasing pressure and fatigue, and provides the elegant, peaceful but superb sensual pleasures for the owner. The concise design of ceiling and natural wood grain have formed the floor of geometric patterns, with the American style's essence. The white marble pillars express a more rational, balanced spirit with the pursuit of freedom and advocating innovation.

富丽的窗帘帷幄和水晶吊灯的搭配为以硬朗线条为主的空间增添了一分柔美浪漫的气氛。主卧及多功能厅天花花纹的使用仿佛云卷云舒的苍穹，让理性的平衡中多了一丝丝神秘。宽大的沙发和椅子透着古朴与沉稳，坐在其中不管是惬意地阅读抑或是沉醉于醇香的红酒，都有让人从酒店或者是办公室中解脱，回归到家中的舒适和随意。湖蓝色和桃红色装饰品的运用为深褐色的沉稳增添了一抹亮色，客厅和餐厅则在锡铅合金烛台、雕花边柜的装饰中，呈现出独特的韵味。

The combination of magnificent curtains crystal chandelier has added a touch of graceful and romantic ambiance into the space. The ceiling patterns in master bedroom and multifunctional room is just like the gathering and spreading clouds, to add a little mystery into the rational balance. The capacious sofa and chairs send out plainness and steadiness. One can get the comfort and leisure as going back home after getting away from the five-star hotel or office while sitting there and either doing some relaxed reading or being intoxicated with the mellow wine. The use of light blue and pink ornaments has added the bright color for the steady dark brown, as the unique charm is revealed on the tinsel candleholders and carving sideboard in the living and dining room.

✸ 青绿山水画
BLUE-GREEN LANDSCAPE PAINTING

青绿山水画是中国画中以浓重的矿物颜料石青和石绿为主的绘画作品，表现山石树木的苍翠而得名。也有在青绿山石的轮廓上勾以金石，又称金碧山水。从六朝开始，逐步发展至唐代二李确立了青绿山水的基本创作特色，两宋之交前后形成金碧山水、大青绿山水、小青绿山水三个门类，在元、明、清三朝各自发展并相互影响，而以小青绿山水为盛。金碧山水重在金碧辉煌，大青绿山水长于灿烂明艳，小青绿山水妙在温蕴俊秀。后者在明末出现，以蓝瑛的没骨重彩山水为代表。

卡尔·拉格斐"蓝色，是天空的色彩，无论是白昼还是黑夜"。绿色是希望，蓝色引领平静，赭褐色调和下的温柔相待，带给生命旅程的又一次华丽复兴。

Karl Lagerfeld has expressed "Blue, is the color of the sky, either at daytime or in night". Green is hope, while blue means for peace, and the softness by ochre brings another gorgeous renaissance for the life journey.

灵感来源分析："顺其自然"，自然界生长的万物必有其存在的道理。生长万物而不据为己有，抚育万物而不自恃有功，以一物导引万物而不主宰，这就是奥妙玄远的德。自从文明的曙光初见，人类就一直把自然作为自己的灵感源泉。以自然界的美好之物结合中国文化作为灵感之源，开始畅想我们的《豪笙印溢》。

The source of inspiration: "Let nature take its course", as all things grown in the nature must have its own meaning of existence. Growing all things without owning them by oneself, tending all things without boasting, leading one thing to another without dominating, this is been the profound morality. Since the arising of civilization, human being has always thinking of the nature as the source of inspiration. We have begun to think about our project with the combination of good things in the nature and the Chinese culture as the source of inspiration.

设计元素的运用：蓝绿色调里的亮面材质无处不彰显着即将破茧而出的蜕变。雾霭一般的灰绿蓝调，深邃的赭褐色温暖营造，让人顿时冷静下来，一种不言而喻的安全感拉开深层次内涵里的优雅序幕。

Use of design elements: The surface texture in the blue-green tone fully shows the great change which will break out. The grayish green and blue tones like the mist, deep ochre have created the warm magic which can make people calm down, while an implied safety has opened the elegant prelude in the deep connotation.

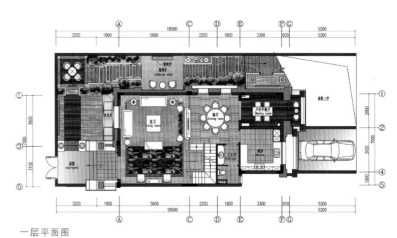

一层平面图
First Floor Plan

地下层平面图

Basement Plan

二层平面图

Second Floor Plan

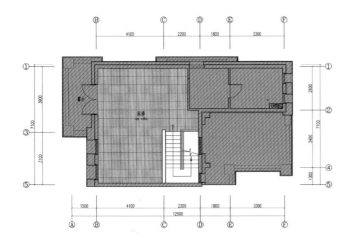

阁楼平面图

Loft Plan

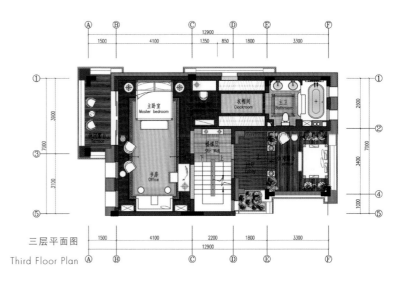

三层平面图

Third Floor Plan

唐乾明月福州接待会所

Clubhouse of
The Villa De Fountain, Fuzhou

名称：

唐乾明月福州接待会所

设计公司：

道和设计机构

设计师：

高雄

摄影：

施凯、李玲玉

主要材料：

黑钛、墙纸、仿古砖、蒙古黑火烧板、绿可木、白色烤漆
玻璃

Name:

Clubhouse of The Villa De Fountain, Fuzhou

Design Company:

DAOHE DESIGN

Designer:

Gao Xiong

Photographer:

Shi Kai, Li Lingyu

Major Materials:

Black steel, wall papers, rustic tile, Mongolian black fire plate,
Lesco wood, white lacquered glass

初入会所，见着空间上的明月造型和软装上相呼应的莲花灯时，脑海中便会浮现出"明月耀清莲"的画面，不禁脱口而出：明月如霜，轻妆照水，纤裳玉立，飘飘似舞。

When you enter the clubhouse, you will immediately see the moon figure and the lotus lamps that correspond to the decorating. Under such environment, an artistic image of a bright moon shining upon purple lotus would suddenly come up in your mind, at that moment, you cannot help but recite those beautiful ancient Chinese poems of the moon.

在会所的入口处树立着若干木桩，有着一种原生态的感动。中式元素的运用，行云流水，一如清莲般恰到好处地出现在属于它的位置上，流畅的视觉引导，自然不做作。只在你伫立或是冥想之时，持一杯清茶就可品得它的万般风情。灯光下，稍作修饰的原木，泛着淡黄色的光泽，一种温暖的感觉在蔓延。眼前的木质纹理显得格外亲近，让人有一种想触摸的冲动，引起人们内心对回归自然的渴望。

Several wood piles casually stand at the entry of the clubhouse, bringing in some primitive atmosphere. The use of Chinese elements in interiors are so natural, as if a lotus appearing right at the spot where it belongs. Under such circumstances, you may taste various flavors by sipping a cup of tea. Under the light, the rustic wood is glowing with pale yellow gloss, providing warmth to the space. And the texture of the wood is displaying directly in front of our eyes, inviting and friendly, and it makes you want to touch those textures by hand and incurs your aspiration of returning to the nature.

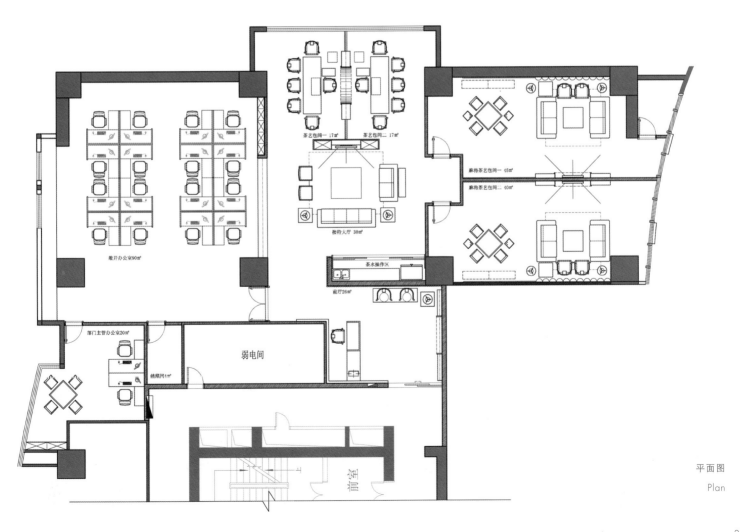

平面图
Plan

※ 旗袍
Mandarin Gown

旗袍起源于满族服装，20世纪上半叶其款式融入西式剪裁而被大幅度改进。旗袍因其代表新时代知识女性的形象而受到欢迎，由上海开始流行至全国，成为民国时期都市妇女的主要服装，并作为世界上影响最大、流传最广的中国传统服装，成为中华民族的传统符号。

走廊的尽头，一个旗袍形状的木架上，挂着一个大红色灯笼。红色在大面积的原木色系的映衬下，少了艳丽之感，印在墙上的红晕，体现出设计者心思的细腻：赋予配饰以思维，用它们的形态绽放设计光彩，延续设计生命，正如那潺潺的流水声，挑动着你的情感起伏。

At the end of the hallway, a big red lantern hangs highly up in the air, upon a wooden shelf in Mandarin Gown shape. The color of red feels less bright and loud setting off by large scale of wood color in the space. And the blush color printed on the wall indicates that instilling the accessories with certain thoughts, which showcases the designer's thoughtfulness and attention to details.

行走于会所内的各个角落，才发现空间在格局划分上采用了半封闭的隔断，不仅实现了视觉上的丰满，气氛上也多了点互动的轻松。有违常规的细节调整，兼顾了会所的私密性与社交性。会所里，半封闭的隔断造型如影随形。透过磨纱材质的玻璃，来往的人影宛若月朗星稀的夜空里，轻覆于皓月表面的那抹柔纱，显得空灵而悠远。

Walking around each corner of the clubhouse, you will find semi-enclosed partitions are applied in the layout of the space. This measure enriches the whole space and adds a bit interacting and leisure atmosphere to the house. These unconventional details and arrangements attentively considerate both privacy and social interactions.

用一种情境的标志，融合在空间的存在美感，造就身临其境的真实。似远非远，似近非近，切身感受着想象的美好，貌似遥不可及，一个转身却又在灯火阑珊处。夹杂在其间的若干绿植陪衬，随人影走动带起花瓣摇摆，轻灵得可爱。掀开垂帘，圈椅上的年轮清晰可见。阳光下的尘埃，就是那似是而非的沉淀。清冷内敛的莲花手座，将禅化解，散落的，是对生活的悟。

A symbol of certain environment may be applied in design to blend in the space to recreate an atmosphere similar to the symbolized environment. Walking casually among this delicately designed space, everything looks far-reaching at one moment, and the next, it just appears right by your side. The plants, the petals, annual rings on the chairs, dust under the sun, the lotus seat, all these small details are all presenting to us the designer's perceptions of life.

用现代手法演绎中式风格的新内涵，颠覆会所的传统概念定位，用标志性的中式元素赋予会所新的空间使命：宁静致远的环境，触手可及的高端，这就是以"新中式主义诗意栖居"为销售亮点的唐乾明月会所。没有醒目的介绍，看不到大张旗鼓的宣传，藏身于市中心的五四路段，可谓是小隐于市。

For this case, the designer applied modern measures to showcase new meaning of Chinese style interiors and completely alter the traditional concept of a clubhouse. The team use some symbolized Chinese elements in the space, and it provides new prospect to the clubhouse – to create a tranquil and relaxing environment that is able to reach within arm's length. This is how New-Chinese style living space is defined. There is no obvious introduction or loud propaganda. It's just a house hidden within a street of the central city, reserved and poetic.

惠州央筑花园洋房样板房设计

Huizhou Central Building
Garden Western-style Showroom

名称：

惠州央筑花园洋房样板房设计

设计公司：

香港 KSL 设计事务所

设计师：

林冠成

主要材料：

橡木、古铜、人造石、黑檀木地板

Name:

Huizhou Central Building Garden Western-style Showroom

Design Company:

KSL DESIGN (HK) LTD.

Designers:

Andy Lam

Major Materials:

oak, bronze, artificial stone, ebony floor

天然木材大量地出现在客厅的设计材料中，木料本身的色彩和纹饰营造了温良敦厚的居家氛围。黑檀木地板的深沉色彩配以顶棚的明亮色调过渡自然而整体和谐。从空间布局到家具造型，硬朗的线条都形成了规整的空间感，墙壁的装饰画则提升了空间意境。

Abundant natural timbers appear on the design materials of living room. The color and the ornamentation of the timbers create the living atmosphere of gentle and unvarnished. The transition from dark color of ebony floors to the bright hue of the ceiling is natural and the whole room is harmonious. From the spatial arrangement to furniture modeling, the hale lines create neat dimensional feeling and the decorative paintings on the wall improve artistic conception of the space.

通过地板色彩的变化成功完成了餐厅、楼梯与客厅的过渡，简洁造型的餐桌选用了深咖色系，一束蓝色的鲜花装点出浪漫的空间氛围。

It finishes the transition from dining room, stairway to the living room by the change of floors' color. The tables with simple shape choose the dark brown color. And a beam fo blue flowers decorate the romantic atmosphere.

楼梯设计最忌暗淡，浅色的地板与璀璨的灯饰瞬间提亮整个空间，通体木制条纹装饰的墙壁则在形成完美分区的同时带来淡雅的自然气息，在灯光的配合下，使空间明亮而清新。

The design of stairways should avoid dimming. The floors of light color and the lighting brighten up the whole space. The wall decorated by wood stripes brings the elegant natural feeling, and it is divided by perfect sections and the space is bright and fresh under the lamplight.

整体色调一致的基础上，卧室的空间氛围更突显温馨与舒适，木纹贴面的墙壁让人心神安宁，或横或竖的几何线条不时出现在顶棚、墙壁与家具上，穿越时空的古韵感扑面而来，舒展而温暖。

The spatial atmosphere of bedroom highlights the warmth and comfort. The wall with wood veneering makes people feel peaceful. The horizontal or vertical geometrical lines appear on the ceiling, wall and furniture. The sense of ancient which passes through the veil of time makes you comfortable and warm.

一层平面图

First Fool Plan

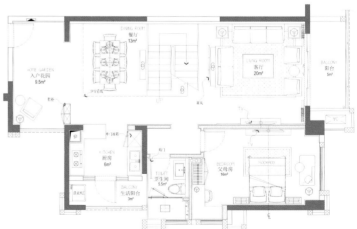

二层平面图

Second Fool Plan

墨绿与条纹是儿童房的活泼元素，从窗帘色调到床品选
择均一丝不苟，清新中带着俏皮，宁静中带着温馨。

Deep green and stripes are the lively elements of
children's room. The choices from the color of curtain
to the quality of bed are meticulous. The whole design
is fresh yet smart, peaceful yet warm.

中洲中央公园样板房

Centralcon Central Park Showroom

名称：

中洲中央公园样板房

设计公司：

香港 KSL 设计事务所

设计师：

林冠成

主要材料：

白砂石、橡木、黑檀木纹、玉石、木地板、烤漆板、墙纸、
皮板

Name:

Centralcon Central Park Showroom

Design Company:

KSL DESIGN (HK) LTD.

Designers:

Andy Lam

Major Materials:

White dinas, Oak, Black Sandal Wood, Jade, Timber floors, Baking
varnish plank , Wallpaper, Dermatome

客厅在色彩搭配上下足了功夫，将黑色的稳重、白色的纯净、藏蓝的冷静与银色的高贵融于整个空间，直线与分割的运用则让空间充满传统理趣。菱格元素用于电视背景和作为空间分割的屏风，墙壁以如橼大笔装饰，配以丝绸抱枕和紫砂茶具，平添了几分文化趣味，使整个空间规整大气，品位高远。

The color combination of the living room takes up much effort of the designer, as the solemnity of black, the purity of white, the calmness of dark navy and the elegance of silver are playing with each other in the whole space. The use of straight lines and segmentation provides the space with traditional charm. Lattice screens are used as TV background wall and partitions of the space. The wall is decorated with traditional writing brush, accessorized with silk cushions and purple clay tea set, which provides the space with some fun and taste.

原木色餐桌保留了明清家具的清雅气度，
藏蓝与白瓷餐具营造出宁静的就餐环境，
一束明黄的春花给沉静的空间增添了一份
活力。餐厅与厨房连通，巧妙地以黑色屏
风和小吧台分隔，妙趣横生。

The wooden dining table reserves its
elegant style of Ming Dynasty and Qing
Dynasty. The color of dark navy and
the color of white from the porcelain
tableware creates a tranquil dining
environment, with a bouquet of bright
yellow flowers flourishing in the middle.
The dining area is connected to the
kitchen, but subtly partitioned by black
screen and mini bar, which is clever and
interesting.

书房以黑茶色为主打色营造了古朴安静的书香氛围，造型简洁的案桌和博古架散发着浓郁的历史文化气息，简洁而有力度的设计提升了空间的格调。

Dark brown is the main colour of the study, for it creates pristine and quiet reading atmosphere. And the simple desk and antique shelf fill the study with rich historic atmosphere. The concise and powerful design promotes the level of space.

卧室色调雅致，空间极为舒展，通体木纹饰面的墙壁搭配黑色菱格装饰板让人心神宁静。飘窗的设计极为精致，两盏古典造型的灯饰仿佛穿越了时空，倚窗而坐，品一杯香茗，享半刻清闲。

The bedroom is quite elegant and spacious. The wood pattern wall covering matches the black lattice decorative board could calm people's minds. The design of the window is also very exquisite, as there are two lamps with classic forms set on the walls.

紧贴墙面设计的收纳柜造型规整、设计精巧，兼具收纳与展示的功能，实现了实用性与装饰性完美结合。

The storage cabinet which abuts against the wall contains the functions of storage and exhibition. It is the perfect combination of practicability and decoration.

平面图
Plan

隽 · 空间

MEANINGFUL SPACE

名称:

隽 · 空间

设计公司:

唐玛国际设计

设计师:

胡建国

摄影:

周跃东

主要材料:

实木地板、仿古砖、马赛克、壁纸、大理石

Name:

MEANINGFUL SPACE

Design Company:

TOMA

Designer:

Hu Jianguo

Photographer:

Zhou Yuedong

Major Materials:

Solid wood flooring, Archaized brick, Mosaic, Wallpaper, Marble

本案为建筑面积约 300 平米的单层住宅空间，空间较宽且采光条件良好，因此设计师围绕这一建筑自身优势以新东方简约主义为主题展开设计，在设计过程中力求达到使用功能与视觉效果之间的平衡，并努力做到古为今用，将中式传统古典建筑元素的精华所在与现代生活的时尚气息融合在一起。

The Project is the single-floor living space with the area of around 300 m². The space is wide and in a good lighting condition, so the designer began to design on the theme of new oriental minimalism with the advantages of this architecture, fully analyzed the equilibrium relationship between function of use and visual effect during the design process, and tried his best to apply the past into the present, to fuse the essence of traditional Chinese classical architectural elements with the fashion in modern life.

将空间造型的比例、尺度、材料质感与色彩搭配和谐地组织在一起，力求能在整个设计中达到合理创新，并能体现传统文化底蕴与新时代人文精神。

The scale, measure, texture are organized harmoniously with color combination to achieve reasonable innovation in the whole design, and reflect the traditional charm and cultural spirit.

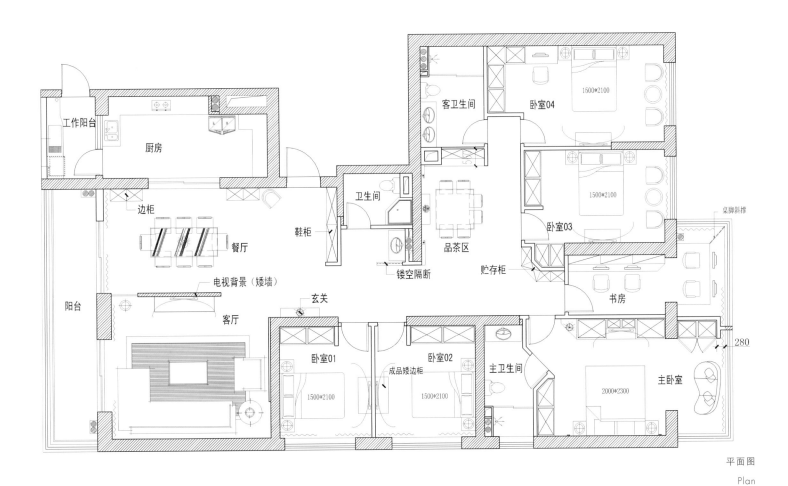

工作阳台

厨房

边柜

餐厅

阳台

电视背景（矮墙）

客厅

鞋柜

玄关

卧室01

1500*2100

成品矮边柜

卧室02

1500*2100

卫生间

镂空隔断

品茶区

主卫生间

贮存柜

书房

桌脚斜撑

客卫生间

卧室04

1500*2100

卧室03

1500*2100

主卧室

2000*2300

280

平面图
Plan

❀ 岩彩画
COLORED ROCK PAINTING

岩彩画，是以宣纸、绢、棉、板、木、壁等为依托物，将五彩的岩石粉末以及金银等金属色媒材以食用明胶或动物胶为粘合剂，定着到画面上的一种绘画。岩彩画做为区别于中国水墨、工笔、重彩、油画、水彩、水粉等其它绘画类别的一种东方式绘画，它注重肌理与造型的对比，欣赏的是一种以平面为主的现代美，强调现代感和时代特征。

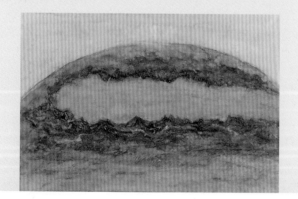

335

图书在版编目（ＣＩＰ）数据

中式空间 / 凤凰空间·华南编辑部编. --南京：
江苏凤凰科学技术出版社，2014.9
ISBN 978-7-5537-3502-3

Ⅰ．①中… Ⅱ．①凤… Ⅲ．①住宅－室内装饰设计
Ⅳ．①TU241

中国版本图书馆CIP数据核字(2014)第156524号

中式空间

编　　　者	凤凰空间·华南编辑部
项 目 策 划	郑　青　宋　君
责 任 编 辑	刘屹立
特 约 编 辑	宋　君

出 版 发 行	凤凰出版传媒股份有限公司
	江苏凤凰科学技术出版社
出版社地址	南京市湖南路1号A楼，邮编：210009
出版社网址	http://www.pspress.cn
总 经 销	天津凤凰空间文化传媒有限公司
总经销网址	http://www.ifengspace.cn
经 　 销	全国新华书店
印 　 刷	利丰雅高印刷（深圳）有限公司

开 　 本	965 mm×1270 mm　1 / 16
印 　 张	21
字 　 数	168 000
版 　 次	2014年9月第1版
印 　 次	2014年9月第1次印刷

标 准 书 号	ISBN 978-7-5537-3502-3
定 　 价	338.00元（USD60.00）（精）